THE NEW NATURALIST

A SURVEY OF BRITISH NATURAL HISTORY

BRITISH PLANT LIFE

The aim of this series is to interest the general reader in the wild life of Britain by recapturing the inquiring spirit of the old naturalists. The Editors believe that the natural pride of the British public in the native fauna and flora, to which must be added concern for their conservation, is best fostered by maintaining a high standard of accuracy combined with clarity of exposition in presenting the results of modern scientific research. The plants and animals are described in relation to their homes and habitats and are portrayed in the full beauty of their natural colours, by the latest methods of colour photography and reproduction.

THE NEW NATURALIST

BRITISH PLANT LIFE

by

W. B. TURRILL

D.Sc. Lond., F.L.S.

KEEPER OF THE HERBARIUM AND LIBRARY
ROYAL BOTANIC GARDENS, KEW

WITH 53 COLOUR PHOTOGRAPHS
BY JOHN MARKHAM, BRIAN PERKINS
F. BALLARD AND OTHERS
27 PHOTOGRAPHS IN BLACK AND WHITE
8 MAPS AND 2 DIAGRAMS

COLLINS ST. JAMES'S PLACE LONDON

First published in 1948 by
Collins 14 St. James's Place London
Produced in conjunction with Adprint
and printed in Great Britain by
Collins Clear-Type Press
London and Glasgow

TO THE MEMORY

OF MY

MOTHER

TO WHOM I OWE MY LOVE

OF PLANTS

CONTENTS

CONTENTS (*continued*)

CONTENTS (*continued*)

COLOUR PLATES

COLOUR PLATES *(continued)*

PLATES IN BLACK AND WHITE

It should be noted that throughout this book Plate numbers in arabic figures refer to Colour Plates, while roman numerals are used for Black-and-White Plates

DIAGRAMS AND MAPS

Every care has been taken by the Editors to ensure the scientific accuracy of factual statements in these volumes, but the sole responsibility for the interpretation of facts rests with the Authors

EDITORS' PREFACE

DURING the last twenty or thirty years a marked change has taken place in our approach to the study of the British Flora. In the early years of this century the work of enumerating the species and their distribution, which had been tirelessly pursued since the days of William Turner in the middle of the sixteenth century, had been virtually completed, except in critical groups. About this time genetics, cytology and ecology were beginning to throw fresh light on how one species has developed into another, and on the infinite complexity of the relation between a plant and its environment ; in short, on the evolutionary biology of plants. This new knowledge has led British botanists to look beyond a bare description of our flora to the evolutionary history of the species comprising it, and the time is now ripe for a survey and assessment of the results of this new approach to British plant life. No botanist has contributed more to these recent advances than Dr. Turrill, Keeper of the Herbarium at the Royal Botanic Gardens, Kew, and in the present volume he has, we feel, written a book which not only gathers up, in a masterly way, the results of previous work, but which will also form a starting point for future researches in the same field.

Dr. Turrill's book breaks new ground in its attempt to present this new material to the general reader and in parts is necessarily somewhat technical. We are convinced, however, that the effort required to grasp the ideas behind the story he tells will be repaid many times by the new vistas that he opens up of field and experimental work at our very doorsteps. Constantly, throughout the book, he suggests fruitful lines of investigation that can be followed by amateur botanists, who may thus be able to add very considerably to scientific knowledge by simple observation and experiment on quite common species.

Dr. Turrill's volume does not add just one more to the already long list of delightful discourses on our wild flowers ; it is a signpost pointing towards a fuller and deeper knowledge of our flora and we hope that many will be encouraged to follow the road to which it points.

THE EDITORS

AUTHOR'S PREFACE

It is, perhaps, of some significance that the study of plant life is almost always referred to as " Botany " and attempts to replace this word by " Phytology " would almost certainly fail. " Ologies " are *parts* of Botany. Thus, there is morphology, the study of form, and physiology, the study of function. Botanists, even professional and academic ones, have rightly insisted again and again that the study of form must not be divorced from the study of function. Botany is, and must remain, the study of all the structures and activities of plant life. Of course, specialization is essential as a method of research but it must be constantly counter-balanced by wide comparative surveys which lead to integration of discoveries, whether of fact or of interpretation. What is true for Botany as a whole is true for the Botany of the British Isles. Intensive and extensive studies have, separately *and* together, the function of advancing knowledge regarding our native and introduced plants. There are many works—" handbooks," county floras, text-books of ecology, and almost innumerable papers in very numerous periodicals—dealing with this or that aspect of British plant life. The present work does not compete with any of these. It is intended as an introduction to the study of British plants as living changing organisms, with a past, a present, and a future. This has meant the linking together of subjects which are most often isolated in published books and papers. The author is himself convinced that much energy is wasted by field naturalists in the study of British plants through undue emphasis being placed upon rarities and " new records." There are so many problems awaiting solution that could be investigated, and many of them solved, by carefully thought-out experiments and patient observations on our common plants that it seems a great pity not to direct attention to the fascinating fields for research provided by " buttercups and daisies."

Very general " terms of reference " were given by the editors of the New Naturalist series; within them the author had no easy task in selection of subjects to be included. On the inclusion side was the desire to maintain a continuous story and yet particularly to emphasize new or relatively new viewpoints. On the exclusion side limitations of space were met by avoiding, within reasonable limits, accounts of special groups or subjects which it was known are to be dealt with in separate volumes of the NEW NATURALIST series. This statement should explain at least some of the obvious sins of omission. It is hoped that mistakes in presentation of facts are few, but the work has been written in intervals of spare time during the past three years under somewhat difficult conditions.

W. B. T.

HERBARIUM HOUSE,
KEW
8 August, 1946.

CHAPTER I

INTRODUCTION TO THE STUDY OF BRITISH PLANTS

AT THE BEGINNING of the Second World War an old Yorkshire botanist, now deceased, wrote for some information. In his letter he referred to the outbreak of war and then wrote, " Mankind is composed of fools, always has been, always will be. I thank God that in the days of my youth my thoughts were turned to the study of plants." We may not agree with this judgment on our own species, or with its logical consequences, but many of us will understand the pleasure and relief of turning from mundane troubles to " consider the lilies." There are no doubt many factors of inheritance and environment which result in the making of a botanist. A desire to know and understand, a capacity for taking infinite pains, persistence in overcoming difficulties, and the power of working through sometimes monotonous preliminaries without losing sight of a main aim are shared with other scientists. The study of living organisms appeals to those who appreciate variety of form, colour, and function, and, above all, the beauty of regular change associated with the interlocking of structure and kinetic function. An important factor, even if it be regarded as less praiseworthy, in the make-up of many botanists is the desire to collect. Psychologists would no doubt classify this trait as a mere example of the acquisitive instinct. One need only say that it is less harmful than some other forms of acquisition even if, like them, it cannot be appeased.

Besides ministering to the aesthetic and acquisitive aspects of human nature the study of plants has the very practical advantages that, in one way or another, it can be carried on almost everywhere, the whole year through, and without great expenditure of money. So much can be done without expensive apparatus or palatial laboratories. Certainly money cannot be " got out of it," but for a low cost a great return in knowledge and in mental and physical health can be confidently expected. In collecting and experimenting the botanist has an advantage over the zoologist in that he does not inflict pain on organisms with a specialized nervous system.

There are many ways in which plants can be studied and it is one of the purposes of this book to emphasize the value of a wide approach. Specialization is unavoidable and, within reasonable limits, is desirable for us all. It has been said that specialization results in " knowing more and more about less and less." In dealing with plant life this is at most only partially true because any investigation concerned with plants sooner or later brings in their surroundings or environment. This immediately widens research and makes it much more exciting. It is, indeed, one of the most attractive features of studying plant life that a very great many questions are invariably opened up by the simplest observation or experiment. It is not very easy to keep a reasonable balance between too close concentration on one main problem and attempting to delve into alluring side issues. One way of maintaining the balance is to plan an investigation on one problem and then gradually to tackle it from as many angles by as many methods as possible. At any rate, if the study of plant life is to give the highest returns in intellectual and aesthetic pleasure, it must not be haphazard but must be thought out first, and a certain elasticity allowed in the scheme of study adopted.

It has frequently been stated that there are two " causes " for plants or animals being what they are and behaving as they do. These two "causes" are often opposed as "nature" versus "nurture" or "heredity" versus " environment." It is convenient sometimes to distinguish effects due, in some major sense, to inheritance from ancestors from those due to the direct action of the environment on the living, and therefore changing, individual. To separate " nature " and " nurture " is, however, a mere artificial, though sometimes useful, device. It is not possible to imagine, relative to living organisms as we know them, heredity apart from environment. That which is inherited from parents must have an environment in which to develop ; " surroundings " must have something to surround, generation after generation.

It is much the same with other aspects of plant or animal life. Structure and function, for example, cannot be divorced in the living organism though in attempting to find out how a plant is constructed we may try to ignore what it does and even to take no account of the special activities of its parts (or organs). We can examine structure in dead material, which may be preserved in spirit or other liquids or dried as herbarium specimens, but behaviour can only be studied in living plants.

PLATE I

F. Ballard

DARK MULLEIN, *Verbascum nigrum* (Scrophulariaceae)
Field near the Herbarium, Kew, Surrey. July 1946

PLATE II

F. Ballard

MILKVETCH, *Astragalus glycyphyllos* (Leguminosae). Bagley Wood, Berkshire. June 1945

The great point is that no one way of studying plants is the only or even the best way. Our minds so work that we can only take in "one thing at a time" but that "thing" may be reduced to the greatest possible simplicity (analysis) or be a built-up structure which is composite but unified (synthesis). Our outlook must be deep and wide in due proportion; our researches must be intensive and extensive.

From the very start of our investigation into plant life there are two methods we can adopt. We can learn much by observation when this is systematic and then more by experimenting when this is properly controlled. Observations and experiments must both be fully and immediately recorded in words and figures and by diagrams, drawings, photographs, and maps. There is no antagonism between observation and experiment and they cannot be sharply separated one from the other. Repeated observations are often closely akin to experiments and it is always a good rule to confirm the results of experiments, as far as possible, by observations of what happens "in nature." One sometimes has the feeling that the very rigid control necessary in many experiments creates such an extremely artificial environment for the subject of the experiment that the behaviour recorded is almost pathological, or at least very different from what happens when the plant is growing under normal conditions. The field naturalist has every reason to claim that his observations, when faithfully recorded, are as valid and can be as important as the results of artificial experiments.

A fascinating study is provided by concentrating on one or a few kinds of plants and learning all that is possible about them by field observations, field experiments, transplanting and testing behaviour under different environments, controlled breeding, and laboratory research. If two or three enthusiasts can form a team the work becomes more stimulating. A very wide field is opened up by using plants themselves as recording instruments. Plants as they grow react differently to variations in temperature, water-supply, and other conditions which together make up their surroundings. If we standardize a number of plants as instruments and learn to "read" their reactions we have a recording apparatus which sums up all the environmental influences which are of importance to the species we use. This is the principle underlying transplant experiments. These consist essentially in growing plants, strictly of one kind, under different conditions. The reactions of the plants are fully recorded and, after the records have been studied comparatively, conclusions, often most illuminating, are drawn from the results. Anyone with a garden can help greatly to increase our knowledge of the behaviour and variations of British plants by simple experiments of this kind.

The chapters that follow deal with many aspects of British plant life and it is hoped they will provide some guide to what is already known. Appeals are made in many places for more research and it is hoped that readers will be encouraged to undertake this. Whether one wants " to learn more about plants " or to sample the exhilaration of discovering about plants facts " new to science," it is essential to go to the plants themselves. Books, lectures, and advice have their place, but they cannot replace actual first-hand contact with living organisms. Get out into the woods and fields, wander along rivers and streams, climb mountains, paddle over salt-marshes, fool about in a garden, as you will, but get amongst the plants and see them, examine them, concentrate on them, and make them answer the hundreds of questions you ask them. All the time you must think and plan. Do not be discouraged by botanical terms or Latin names. They are not difficult to understand if taken a few at a time. They have, or should have, the great advantage of precision—of enabling a great deal to be said exactly in few words. Some " jargon " is essential, though it can, of course, be overdone. It is worth while to master such terms as are used in this book since many of them are now being used rather widely even in " popular " print, wireless talks, and films. Biology is the most delightful of all subjects but it is not the easiest to study. Living organisms are exceedingly numerous, very varied, and anything but static. Botany and zoology are, therefore, fundamentally more difficult subjects than chemistry and physics. For reasons that are not entirely creditable to mankind, chemistry and physics have made much greater progress than has biology in the last two centuries. Progress in some branches of biology has to wait upon advances in chemistry and physics and the peace and happiness of the world would be less in danger if chemists and physicists would turn their attention to biological problems rather than to explosives and bombs, atomic or otherwise.

The complexity of living organisms and their behaviour is one reason why in biology it is possible to make very few statements which have the status of " natural laws." Most broad generalizations have exceptions. We say that plants and organisms occur as " individuals " classifiable into " species " but it is not possible to give definitions which exactly delimit " individuals " or " species "—working definitions are very useful but " doubtful individuals " and " critical species " are always with us. Biologists nowadays accept the theory of evolution and it is used in the following chapters to explain many of the facts recorded for the British flora but evolution has been by different mechanisms, working at diverse rates at different times and in different places. Perhaps the " cell theory "—that all plants and animals are built up of one or more units of living substance, the units being termed " cells "—is the nearest approach to a " law " we have in biology. One has, however, to acknowledge that the units are sometimes so inseparable that to call them " cells " approaches absurdity, as in the threads of some Algae and Fungi.

Our British flora is not a large one but in it are representatives of all the great groups so that wild material is available for the study of many kinds of botanical problems. It is essential to determine accurately any plants studied. There are various books (often called " Floras ") with descriptions, keys, and sometimes illustrations which have been prepared specially for " naming " plants. Determining plants leads on to the many problems of plant classification (or taxonomy). This is a subject very attractive to the logical, methodical, and collector-type of botanist. While this book is not primarily concerned with classification the importance of the subject is fully realized and much in the following chapters depends on and, in return, can be used in classification.

The study of the structure of plants (morphology) is of less interest by itself than when combined with consideration of function or behaviour (physiology). For the British flora this is particularly true since we have few species of eccentric form and the general structure of many of our plants is now fairly well understood. On the other hand both the relationship between form and habitat and the bases of inheritance of structures necessitate very careful examination of organs and their characters and there is a very great deal of detailed research needed on British plants along these lines. The study of the "home-life" of single species (or of individuals) is termed "autecology" and the study of vegetation, of plants grouped in communities irrespective of their systematic relationships, is called in contrast "synecology." Attempts to penetrate the mysteries of heredity by modern methods of research come under the heading of " genetics." Since genetics is now very closely linked, both in theory and practice, with " cytology " or the study of the minute details of cells and particularly of the nucleus (the specialized portion of living substance which largely controls cell activities and also carries the hereditary factors) the term "cytogenetics" (cytology plus genetics) has become fashionable. There is no doubt that the most favourable opportunities for exploring new ground in the realm of British plant life are provided by combining the methods of ecology and cytogenetics in investigation of the one set of problems on the same (or equivalent) material. This synthetic approach has many advantages: it combines intensive and extensive outlooks as well as observation and experiment; it gives plenty of opportunity for research in the field, in the experimental ground or garden, and in the laboratory or herbarium; it links "nature" and "nurture"; and it allows for team work. What more can the naturalist want? Two requisites are essential for success: utmost care in observing and experimenting, including record-keeping; and infinite patience and persistence. Given these, the investigator is bound to make worth-while discoveries which will be the more important in proportion to the ingenuity with which the scheme of observations and experiments has been thought out beforehand. It is hoped that the following chapters, with all their imperfections of contents and presentation, will introduce many readers to the joys of knowing something about British plant life, will encourage others to find out more, and will stimulate some to concentrate on investigations which will make what we know, or think we know, speedily out of date.

CHAPTER 2

THE VERY BEGINNING

OUR EARTH is one of a number of planets revolving round the sun. Its exact mode of origin is still largely a matter of speculation but astronomers and geologists appear to be agreed that at one time the nature and temperature of its surface or periphery were such that no plants or animals approximating in structure and behaviour, even remotely, to those now existing or to those whose remains we know as fossils, could exist on our globe. There are thus only two ultimate origins that are possible for the living organisms composing past and present floras and faunas: after solidification and cooling of its surface living organisms either emigrated to the earth from some outside source or originated on the earth from non-living matter.

The enormous number of stars, estimated at a hundred thousand million, in our own star system or "island galaxy," and the enormous number of other star systems, again estimated at a hundred thousand million, make it very possible, if not probable, that the earth is not unique in the cosmos as a habitation for living organisms. There are, however, many factors which make it most unlikely that habitations suitable for plants and animals, at all similar to those we know here, are very frequent relative to the number of "heavenly bodies." Further, the inconceivable distances that separate the stars and their planetary systems one from another, together with the physical conditions of the intervening space, make it most unlikely, if not impossible, that any form of living organism could migrate from star to star. Such a statement is still more applicable to the galaxies.

There remain the other planets of our solar system. It appears from the most recent researches and conclusions of astronomers that only Venus and Mars of the sun's family, outside the earth, have conditions that could sustain living organisms. Plants must have, at least, oxygen, water, certain "mineral" food substances, a narrow range of temperature, and, with some exceptions, light, in order to grow and reproduce. Venus has probably a humid and hot surface but no or very little oxygen in its atmosphere. It may be the planet of the future, in the sense that when its surface becomes cooler living organisms may appear and evolve, at first under hot swampy conditions. Mars, on the other hand, belongs to the past. The reddish colour of its surface appears to be due to the oxidation of iron, and this points to the presence of both oxygen and carbon dioxide. Mars, however, has probably lost most of its oxygen and though some of its surface phenomena have been interpreted as signs of vegetation other explanations of them are possible and the occurrence of animal life on its surface is purely speculative. Mars, when nearest to the earth, is about 36,000,000 miles distant. This great distance, though of a totally different order from that of the nearest " fixed " star to the earth (some 26 million million miles), makes one hesitate to assume that any living organisms, even in the form of the smallest germs, could have reached the earth from Mars and still less from any other planet. It is true that the spores of certain Bacteria and other plants, and even the seeds of some flowering plants, can withstand very great extremes of temperature, drought and other environmental conditions, and some very small organisms and parts of organisms have been found in the upper layers of the earth's atmosphere. Nevertheless, the great cold of interplanetary space and the long time-interval which even the shortest or most direct route would involve, even if such was accidentally used—and the chances against its use are enormous—are decidedly not in favour of immigration of organisms from outside the earth. The only supposed positive evidence is the reported occurrence of Bacteria in stones or pieces of matter which have fallen through the atmosphere (aerolites) but it remains doubtful if these came originally from other planets. If they did, it still remains very doubtful whether such Bacteria could be regarded as ancestors to the plants and animals now found on the earth. Even if one does assume that organisms originated outside the earth and subsequently immigrated, this merely pushes back their origin in space and probably time and there seems no reason why if they have originated elsewhere they have not also originated on the earth.

Plants and animals are here on our earth, and the evidence from their fossil remains in the sedimentary rocks of the earth's crust is conclusive that they have existed as separate groups for many millions of years. If, then, we reject as unlikely that the original organism or organisms reached the earth from outside we are left to consider their origin on our own planet. Though this too is a highly speculative question a considerable number of facts relevant to possible answers are known.

A living plant or animal is obviously different from a dead plant or animal of the same kind, yet none knows fully and exactly how being alive differs from being dead. The bodies of living organisms are composed of chemical elements and a majority of the 92 elements have at one time or another been found in plants or animals, though it does not follow that all those found are essential to living. Such elements as carbon, hydrogen, oxygen, nitrogen, phosphorus, sulphur, potassium, calcium, and a few others are constantly found in tissues of both plants and animals and their importance, as components of essential structures, cannot be questioned. Actually the elements are present as chemical compounds often of great complexity. Thus protoplasm, the jelly-like material of living cells which has been termed " the physical basis of life," is chemically composed of proteins which are themselves complexes of other compounds known as amino-acids and composed of carbon, hydrogen, oxygen, nitrogen, and sometimes sulphur and phosphorus also. A point of major importance is that the chemistry of living organisms is based essentially on the element carbon and its peculiarities. The only known exceptions to this statement are certain bacteria which utilize—partially or entirely in place of carbon—iron, sulphur, or hydrogen as a source of energy. Organic chemistry is essentially the chemistry of carbon compounds and, though many such organic compounds are now made in the laboratory, in nature the more complicated carbon compounds are all made, so far as we know, only by and in plants and animals. One can, indeed, accept a further statement, that " all flesh is grass and all fish diatom," in so far as this implies that food manufacture can be traced back ultimately to the action of the green matter of such plants as grasses and diatoms, in utilizing the energy of the sun's rays to build up sugar from carbon dioxide and water. It is right to mention diatoms since these microscopic plants, with often beautiful skeletons of silica, form with some other minute green organisms the sole final source of food in the seas. Every food cycle can be traced back to green plants—a lion eats a zebra which eats herbs, and however long a chain of one organism living on another may be unravelled, the activity of the green colouring matter of plants has to be brought in before a basis is reached in the manufacture of sugars, starches, etc. (the carbohydrates of the biochemist) from inorganic raw materials. This does not, of course, mean that simple inorganic substances are not also used by the plant or animal body in important or even essential functions. It does, however, mean that very early in the evolution of plants and animals, and perhaps almost at the beginning, photosynthesis (as this process of sugar manufacture by the green colouring matter or chlorophyll is termed) existed. Many speculations on " the origin of life " appear to ignore this fact.

While, then, living organisms have material bodies differing from inorganic matter chiefly in the complexity of the component compounds this is no absolute difference, the more so that the chemist can now manufacture a host of similar compounds from inorganic substances, and ultimately from elements. It is, moreover, well within the range of possibility that conditions, such as great heat, electrical discharges, concurrence of certain substances, mass predominance of this or that substance, and so on, used by the chemist in his laboratory occurred naturally on the earth's surface with considerable frequency in the early history of the earth as a solid or solidifying body. We must turn, therefore, to the activities of living organisms and enquire if they throw light on the nature of living.

Growth, maintenance, and reproduction are obvious features of life-cycles. It is usually undesirable to introduce any ideas of conscious purpose into the study of plant life but it is often helpful to remember that most functions performed by organisms can be classified either as those necessary or advantageous to the individual (or sometimes colony) or those resulting in reproduction or some other form of multiplication. Even the poet Tennyson (IN MEMORIAM LV, v. 2) partly corrects his somewhat dogmatic statement regarding " Nature " that:

> So careful of the type she seems,
> So careless of the single life.

One may add that, metaphorically speaking, " Nature " does a great deal towards taking care of the individual, since the " type " is composed of and depends on " single lives." The story of growth, from acorn to mature oak, for example, records a series of reactions between the germ, seedling, sapling, or tree on the one hand and the environment, the whole complex of ever-changing surroundings, on the other, which would require a volume to describe. It is true that some of these reactions can, within limits, be mimicked with inorganic materials and the use of controlled forms of energy in the laboratory. Experiments along these lines are not without interest as they sometimes suggest the physico-chemical mechanisms involved in this or that

PLATE III

Brian Perkins

OXLIP, *Primula elatior* auct. (Primulaceae). Near Stetchworth, Cambridgeshire. May 1943

Robert Adam

WOOLLY WILLOW, *Salix lanata* (Salicaceae). Clova Hills, Forfarshire. July 1937

happening in the living organism. They are, however, at best, similarities or analogies which sooner or later break down. The chemist has not yet made " living protoplasm " ; the physiologist has not yet made a " living organism." A living organism is in some ways more than the sum of its parts and this fact is recorded in the statement that any complete explanation of it in terms of a machine or a physico-chemical robot would have to include the properties of self-feeding, self-regulating, self-adapting, and self-reproducing. Moreover, these, and other phenomena, express themselves within a generally unified life-history, though apparent antagonisms may exceptionally be noted, as when an over-fed plant does not reproduce by flowering and seed-setting. It is most important to bear in mind that all the varied activities of an organism have to work together and a balance has to be kept between them. We shall meet with striking examples later on in our study of adaptations of British plants to their surroundings, and here one example must suffice. Plants must have water. This, in the flowering plants, with a few exceptions, is obtained from the soil *via* the root hairs and roots. It is, however, not merely the absolute total of water taken in that is important but the balance between the water absorbed and that lost in the giving off of water vapour by the aerial parts, mainly the leaves and shoots.

We are thus forced to the conclusion that living organisms have bodies of a physico-chemical make-up, much more complex than anything found in the natural inorganic world but not necessarily otherwise different from inorganic materials. In addition, they perform functions which are harmoniously balanced so as to allow the organism both to live its life and to reproduce its kind. Are physico-chemical phenomena alone sufficient to account for the making and maintenance of living organisms and their reproduction or is something else, some vital principle, involved ? There are two main schools of thought on this question. They are not entirely uniform and are linked by modifications and compromises, but it will serve for our immediate purpose to note that those who accept the hypothesis that " life " is something distinct from matter, something added to or working in and with matter in a living organism, are termed vitalists, while those who hold that the properties of matter are sufficient to explain all the phenomena associated with plants and animals are called materialists, or mechanists or, best, non-vitalists. If " life " be merely one of the characters of certain kinds of matter it follows that matter is far more wonderful than the physicist and chemist have yet shown it to be since it must include all the phenomena of living. The naturalist can regard such possibilities as tools which, like all tools, have wider or narrower uses and limitations. There is no doubt that the physico-chemical tool is, at the present state of our knowledge, very useful, yet it is obvious that, in some sense, living organisms are different from and behave differently from non-living matter. Something is always lost in analysis. The first thing a chemist does when he analyses " living matter " is to kill it and though the physiologist may concern himself with the doings of plants and animals in the live condition he cannot pass from them to the physico-chemical background of their behaviour except by jumping a gap.

It is obvious that in the preceding paragraphs we have touched, and in part passed beyond, the bounds of knowledge based on scientifically determined facts. There is also a great element of speculation in the answers to the problem to which we now return : that of the origin of living organisms on the earth's surface. As we have seen, it is more likely that plants and animals originated on earth than that they reached our globe, even in some common ancestral form, from other planets or star systems. The evidence in favour of the evolution of all kinds of organisms so far known from pre-existing organisms is so great that the biologist must accept it as conclusive. Hence the problem narrows down to how one or a few unknown primitive kinds of organisms arose from non-living matter. It is difficult to see anything more remarkable or miraculous in this than in the process of evolution itself.

In many ways, though not in all, animals are more highly evolved than plants. This may be summed up by saying that animals have a greater control over their environment than have plants, and the " higher " animals, mammals especially, to quite a considerable degree, create their own environments. In food-making, too, we remember that plants constitute the link between animals and the inorganic world. Nevertheless, the evidence is all against the course of evolution having been the appearance and development of primitive plant life, the evolution of this, and from a highly developed plant the evolution of a primitive animal. There is little doubt that what we now distinguish as plants and animals evolved from a common stock of primitive organisms with a mixture of plant and animal characters. Many such organisms still exist and a convenient group term for them is "Protista." From Protista, in the broadest sense, it would appear plants and animals evolved with increasing differentiation, that is, by divergent evolution. The wrong (*a*) and the more nearly correct (*b*) viewpoints can be expressed diagrammatically in the following way :

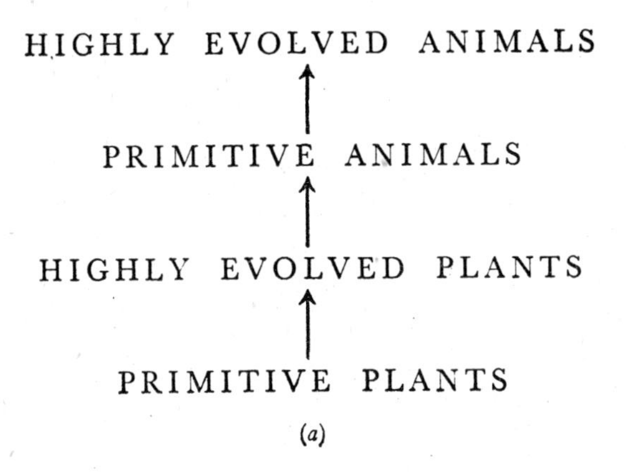

(*a*)

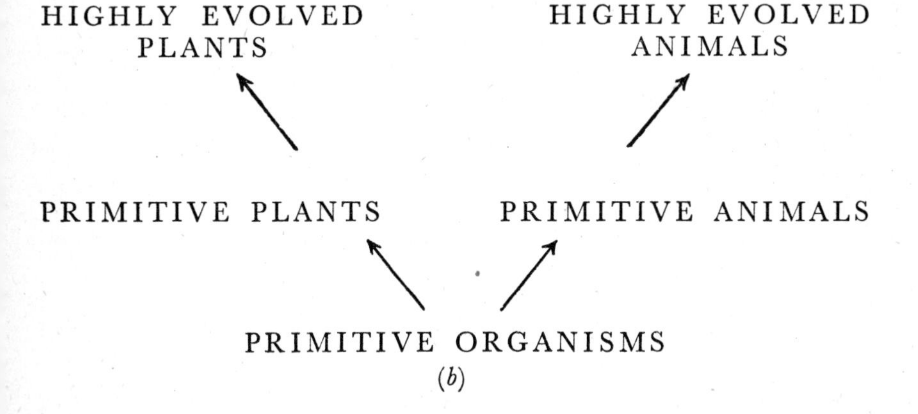

(*b*)

Diagram (*a*) is certainly incorrect, (*b*) is greatly over-simplified but within limits correctly expresses the divergent evolution of the two great groups of living organisms.

Another series of recent discoveries relevant to our present subject concerns what are known as viruses. There are many diseases, often highly infectious, of both plants and animals that are caused by agents of such small size that they pass through the pores of fine filters and cannot be seen directly under even the most powerful microscope. There is, however, a relatively wide range of size between the smallest viruses (such as those of tobacco-mosaics or foot-and-mouth disease) and those which (like the virus of parrot disease) are just resolvable under the compound microscope and thus, as regards size, link on to the smallest known Bacteria. So far as is known the viruses continue their existence as such only in the cells of plants or animals. It has not been found possible to cultivate them on artificial media. There has been much controversy as to whether they are living structures, in the sense that Bacteria are, or non-living chemical bodies. Very recently a few viruses (e.g. the virus of tobacco-mosaic) have been obtained in a pure chemical form by extracting them from the tissues of diseased hosts and such extracted and purified samples have infected new hosts. Such purified yet infectious virus material was certainly not living in the usual sense of the term. Yet viruses can multiply with great rapidity, though only in the tissues of a suitable host, and further show " adaptation " in that they can be greatly modified in the effects they produce by being passed through a series of hosts. Reproduction and adaptation are characteristics of living organisms. Do viruses form a link, or one of several links, between the living and the non-living? The study of viruses is of very great practical importance since they are the causative agents of various devastating diseases of plants and animals, including man, especially under conditions of cultivation or domestication. It would appear also that they have very great theoretical importance since they may well help to solve " the riddle of life." Research on them, from many different angles, is being actively pursued in this and other countries and its results can be awaited with interest by all of us.

Only a passing reference need be made to agents which attack and destroy living cultures of Bacteria, the bacteriophages as they are called. They are smaller than most viruses. It is probable that they are of chemical nature but though many facts regarding their activities are known their exact nature and mode of origin remain uncertain.

The simplest definitely living organisms are the Schizophyta, or splitting-plants, and the Protista which we have already mentioned. The former consist of two main groups, the Bacteria and the blue-green Algae. They are classified as plants because of their methods of spore-formation which are similar in many respects to those found in other groups of undoubted plants. The Bacteria are numerically an enormous assemblage of one-celled organisms found under the most diverse conditions on the land, in the sea, and in the air. The activities of various species include the assimilation of nitrogen (that is, the formation from the free nitrogen of the atmosphere of soluble nitrates that can be used by higher plants), production of some diseases in plants and many more in animals, and the breaking-down of complex organic matter into simpler compounds. The last activity, in the form of putrefaction, is of supreme importance since without it the earth's surface would long ago have been choked up by the dead remains of plants and animals, if, indeed, these could have first existed without Bacteria. It is of interest, too, to recall here that there are some groups of Bacteria obtaining energy by chemical, but not photochemical, means, by the breaking-down of compounds of iron, sulphur or hydrogen without the use of energy supplied by sunlight. They are thus the only organisms completely independent of carbon compounds as a source of energy and energy storage. In other words, they are outside the sunlight–carbon-dioxide–water–chlorophyll scheme of food manufacture. The vast majority of Bacteria, however, depend upon food already made by other living organisms which they obtain as parasites from living hosts or from dead animal and plant remains. To this important extent they are dependent on other organisms and could not have preceded them in time. With a few possible exceptions the special green colouring matter, chlorophyll, does not occur in Bacteria.

The blue-green Algae (Cyanophyceae or Myxophyceae) are, in regard to habitats, of more restricted occurrence than Bacteria but like the latter are found in all parts of the world. They are mainly water organisms, some occurring in fresh and some in salt water, while a number are terrestrial. Of particular interest is the presence of blue-green Algae in hot springs up to temperatures of 75° C. These hot-spring or thermal species are limited in number but some of them have a practically world-wide distribution. Some blue-green Algae are single-celled organisms but in a great many the cells remain united in filaments or in non-filamentous colonies. A general character of the group is the presence of a blue-green pigment, known as phycocyanin. This is present, often with one or more other pigments, together with green chlorophyll in the outer portion of every cell. The blue-green and other colouring matters usually mask the green chlorophyll but the latter enables the plants to manufacture their own food in the usual manner of all green plants.

There are some characters which the Bacteria and blue-green Algae possess in common and of these the following are important : structure of the cell, methods of multiplication, and methods of reproduction. In every cell of the vast majority of plants and animals there is, besides the general jelly-like living protoplasm (often called cytoplasm), at least one especially differentiated body known as a nucleus (plural, nuclei) which is of the utmost significance as a controller of cell activities and a carrier of hereditary factors. The splitting-plants have no definite nuclei separated from the rest of the protoplasm by membranes. The common method of multiplication is by the splitting of one cell into two and repetition of the process. Reproduction is by means of asexual spores of one kind or another. These spores are formed from vegetative cells without the production and fusion of sexual cells. In other words, normal sexuality is absent and this may well be connected with the absence of definite nuclei. It is these peculiarities in structure and behaviour which make it very doubtful if the Bacteria and blue-green Algae can be considered ancestral to the main mass of plants, and still less of animals, which have evolved on the earth and in the waters thereof. Their importance in the "economy of nature" cannot be gainsaid but in origin and relationships they are probably an isolated group of an origin (or origins) independent of that (or those) of other organisms to none of which are they likely to have given rise.

It is then to the large and varied group of " plant-animals " known as the Protista that we turn for suggestions of the most primitive forms of living organisms, such as may conceivably have arisen from inorganic matter. The Protista are small one-celled organisms with a mixture, in varying degrees, of characteristics usually regarded as those of plants and those of animals. As a group they have nuclei and show sexual methods of reproduction. Apparently one of the simplest is "Amoeba," found in water and watery mud. Essentially, it is a translucent mass of slime-like living protoplasm of irregular and constantly changing shape. It moves along with a creeping movement, sending out projections of its protoplasm and drawing the rest of the mass after them. The organism feeds by ingesting small Algae and other minute organisms. It does not manufacture food directly and solely from inorganic materials. Under environmental conditions favourable for growth it multiplies by binary fission, one organism dividing into two, but some kinds at least can reproduce by a relatively simple sexual process. Under certain circumstances the body rounds itself off and secretes a tough case or cyst ; it can then lie dormant and survive both drought and freezing. There is also sometimes a kind of spore formation. The protoplasm of the Amoeba is irritable (i.e. it reacts to stimuli such as an electric shock or a prick) and stimuli are conducted through the protoplasm. In addition, the organism also shows some automatic reactions which cannot be traced to external stimuli. It is evident that even if the Amoeba be the most, or one of the most, primitive of existing organisms there must be a very considerable gap between it and the first form of life that appeared from inorganic matter. What evolutionary stages filled in the gap can only be conjectured. Other Protista have more complicated structures than Amoeba and from some of them possible lines of evolution leading to various plant and animal groups can be suggested.

The problem of how living organisms may have arisen, or may still arise, from inorganic matter remains unsolved. There is no direct and certain evidence by which to solve it and claims to have made living organisms from such inorganic substances as colloidal silica and other colloidal materials are not generally accepted. It was in 1860 that Pasteur showed clearly that living organisms did not arise spontaneously in putrefying matter when this is properly sterilized and the entrance of germs from outside is prevented. The suggestion that Bacteria and blue-green Algae may arise in the chemically rich waters of hot springs at high temperatures is not very convincing since they are well-defined organisms and it is doubtful if forms intermediate between them and inorganic matter have been found. Moreover, it is very unlikely that these groups were the ancestors of the main mass of plants and animals, though they may be a by-product of the early stages of evolution of these last. Most lines of speculation, in so far as they are based on known facts, point to the first occurrence of living organisms in water, it is often said in the sea, but it does not follow that the sea water, when organisms first appeared, had the composition it has now, and even this is not everywhere the same.

Perhaps the most reasonable line of suggestion (it is nothing more) is somewhat as follows. Some time after its origin our planet had cooled down sufficiently to form a kind of crust on which water condensed gradually. The atmosphere probably contained little oxygen but abundance of carbon dioxide. Much of the nitrogen probably became combined in the earth's crust so that ammonia could be formed by the action of water. Ultra-violet rays from the sun could thus reach a mixture of water, carbon dioxide, and ammonia and produce a great variety of " organic " substances, including perhaps sugars and relatively simple nitrogen-containing compounds which could form " food " for the first more or less living things. These latter were possibly large molecules built up under the sun's radiation and, like the existing viruses, only capable of multiplication in a very special environment. By their union and interaction in combination with other more or less complex materials in the chemically rich primeval waters they may have built up some compounds approximating to protoplasm in which reactions occurred and in which the characteristics of " living " emerged.

This speculation, even if true, is insufficient by itself. Though it is possible that the first living organisms found plenty of certain kinds of " food " to hand, we have seen that the existing realm of nature with all its varieties of food cycles is founded on the food-manufacturing power of the colouring matter of green plants. Now this green colouring matter or chlorophyll is by no means a simple chemical substance. It occurs in two slightly different forms, having the empirical formulae $C_{55}H_{72}O_5N_4Mg$ (chlorophyll *a*) and $C_{55}H_{70}O_6N_4Mg$ (chlorophyll *b*) respectively ! The chance formation of a substance with such a complex make-up is almost as difficult to imagine as that of living protoplasm. It is least improbable that the first living organisms lacked chlorophyll and that in due course this was made by the activity of protoplasm. Yet chlorophyll certainly appeared or emerged before evolution of any of the " higher " groups of plants or animals could begin or at least proceed far. From various groups of Protista, some of which contain chlorophyll, it is possible to trace what may reasonably be lines of evolution to groups of Algae and perhaps other plant groups and, along other lines, to trace the gradual differentiation of animal groups.

The tremendous importance of the green chlorophyll and its function of food-manufacture by the capture of the energy of sunlight has already been stressed. In a later chapter (Chapter 12) we shall discuss actual examples of a rather curious general fact. The power of producing chlorophyll must have arisen very early in the evolutionary history of all the main groups of living organisms, yet it is frequently lost in even the most highly evolved plants. Again and again, in breeding research, it is found that " albinos " appear in families derived from green (self-feeding or autotrophic) parents. Such albino seedlings are generally yellowish-white with no tinge of green. Since they soon die (they have been occasionally kept alive artificially by grafting on to a green stock) they do not affect the general population of the species. It is, however, suggestive that a function basic to the food supply of nearly all living organisms can be lost at a single step by a sudden isolated change (or " mutation "). It is not always realized what a " touch and go " business it is to be alive !

From now on, through the remainder of this book, we are concerned only with problems of plant life and, in the main, have to concentrate on the plant kinds which occur or have occurred in the British Isles.

CHAPTER 3

THE RISE OF LAND PLANTS

THE LAST CHAPTER linked up our subject with astronomy, chemistry, and physics. This and the two immediately following ones have connections with geology. As is generally known, geologists by detailed comparative studies of the remains of animals and plants found as fossils in the sedimentary rocks, i.e. those laid down under seas and lakes, have been able to classify these rocks according to their relative age. More recently still, by the study of radio-active minerals in rocks of known relative age, it has been possible to give what may be fairly approximate ages in years to the major divisions. We need not go into further details at the moment but we must have a clear concept of the geological series and understand what is implied by at least a minimum number of names applied to the main divisions. The following table (pp. 22, 23) has, therefore, been prepared in such a manner that it should serve both as a guide to geological terms used in the text and, to a certain extent, as a convenient summary of the more important facts discussed in this and the following chapters.

It should bc cxplained that geologists have, most often, found it convenient to classify the sedimentary rocks by their animal and not by their plant fossils. This is because the fossil remains of animals are much more numerous in the majority of such rocks than are those of plants and for other reasons also are more suitable for purposes of "zonation." One consequence, unfortunate for the botanist, is that the divisions now established, while coinciding with important phases in animal evolution, do not always correspond with those of plant evolution. For example, the modern type of flowering plant first appears, at least abundantly, in fossil form in the middle of one of the great divisions into which the rock strata are classified and before the end of that epoch had gained a dominant position in the world's flora.

Readers unfamiliar with the major groupings of plants may find the outline given in Appendix B useful for reference in connection with this chapter. Much information on the geological strata, their characteristics and outcrops in the British Isles, is given in L. Dudley Stamp's *Britain's Structure and Scenery*, 1946, in the NEW NATURALIST series.

Pre-Cambrian Plants

Though the study of fossil plants tells us a great deal about the changes in flora and vegetation of the earth and, in particular, of what are now the British Isles, it tells us nothing about the origin of plant life and little on which we can rely regarding the plants that existed before the advent of a land flora. The enormous series of rocks, which are the record of a period of time as long as or longer than that covered by all the other geological eras together, known as Pre-Cambrian or Archaean rocks, are devoid or almost devoid of fossils. Certain problematical structures, possibly of plant origin, have been described as supposed Bacteria or Algae from Pre-Cambrian strata (Algonkian) in America and elsewhere. There can be little doubt that precursors of the rich fauna and flora of Palaeozoic times existed when the later Pre-Cambrian rocks originated and it may well be that further research with improved methods will make their satisfactory determination possible. The lack of evidence regarding the earliest plant life is, however, not surprising. Water-plants of to-day have few hard parts which are suitable for preservation as fossils and the evidence is strong that plant life originated in water. One could hardly expect delicate organisms, such as Amoeba or delicate unicellular Algae, to be preserved as fossils in these very old Pre-Cambrian rocks.

Early Palaeozoic Plants

Much of what has been said in the previous paragraph applies also to the rocks of Cambrian, Ordovician, and Silurian ages. There is, however, one very important difference between Pre-Cambrian and early Palaeozoic times. In the latter there was an abundant and varied animal life whose remains could be and were preserved as fossils. These animals must have fed, directly or indirectly, on plants which must have been present in great numbers. During these early epochs, primitive seaweed types were probably dominant. Many impressions and even remains showing apparent cell structure have been described and named as early Palaeozoic Algae. Some of these were capable of secreting lime and were thus important as rock-builders.

It is only when we come to rocks of Upper Silurian age that we have anywhere in the world even moderately satisfactory evidence of the

Era	Major divisions (*epochs*)	Sub-divisions (*stages*)	Commencement of stages (millions of years ago)
CAINOZOIC	QUATERNARY	Present day	—
		Post-glacial	0.01
		Glacial	1
	TERTIARY	Pliocene	15
		Miocene	30
		Oligocene	40
		Eocene	50
MESOZOIC	CRETACEOUS	Upper Cretaceous	90
		Lower Cretaceous	110
	JURASSIC	Upper Jurassic	
		Middle Jurassic	
		Lower Jurassic	150
	RHAETIC		
	TRIASSIC		190
PALAEOZOIC	PERMIAN		225
	CARBONIFEROUS	Upper Carboniferous	
		Lower Carboniferous	240
	DEVONIAN	Upper Devonian	
		Middle Devonian	
		Lower Devonian	250
	SILURIAN		300
	ORDOVICIAN		340
	CAMBRIAN	Upper Cambrian	
		Lower Cambrian	480
AGNOSTOZOIC	PRE-CAMBRIAN OR ARCHAEAN	Late Pre-Cambrian	800
	BEGINNING OF THE EARTH		not less than 1,900–3,000

Botanical characters and other remarks
Flora as now existing and as it has existed within historic time. Peat deposits, submerged forests, river-gravels, etc. Various glacial and inter-glacial deposits, some with plant remains of a depauperated flora. Includes the Cromer forest-bed and East Anglian Crags. Flora temperate. At most, little represented in Britain. Alps and other mountains raised. Includes plant-beds of Bovey Tracey (Devonshire), Bembridge (Isle of Wight), etc. Includes the London Clay and numerous fossil plants such as those at Alum Bay (Isle of Wight), Bournemouth, Barton, etc. Flora tropical.
With Upper Greensand and Chalk. Flora dominated by flowering plants (Angiospermae) of more or less tropical affinities. With Lower Greensand and Wealden. In the former flowering plants (Angiospermae) have been found as fossils but not in the latter. Some of the beds, especially those of Jurassic age, with rich deposits of fossil plants, many of them seed-bearing but allied to cycads, monkey-puzzle, pines, etc. (Gymnospermae). Also many ferns and other vascular cryptogams.
A rather depauperated flora of Carboniferous type. Includes the Coal Measures with the richest of all fossil plant deposits in Britain. The flora quite different from any now existing and dominated by extinct groups of Gymnospermae and groups of vascular cryptogams now either extinct or represented by kinds very different in habit. With the Millstone Grit and Carboniferous Limestone. The best fossil plant remains in Britain are in southern Scotland. Flora obviously ancestral to that of the Carboniferous epoch. Includes Rhynie chert plant-beds of Scotland. Flora with early types of land plants. The first known land plants. Plant remains often doubtful but suggest aquatic modes of life only.
Plant remains of any sort very doubtful.

The estimates, in millions of years, for the commencement of periods in this table (as given in column 4) are based on recent evidence, but may have to be very considerably corrected if new data on this difficult subject be obtained.

existence of land plants, especially from Victoria, Australia. These remains do not tell us very much about the habit and structure of the first land plants, and it is only in rocks of the next epoch that fossil plants have been found in such a state of fossilization that relatively full information has been obtained about them and the best of these happen to be British.

Devonian Plants

Many remains of land or marsh plants of Devonian age have been recorded from the British Isles and elsewhere but for the most part they are either scrappy or badly preserved. In 1913, however, there were discovered near Rhynie, Aberdeenshire, petrified plant materials of Middle Devonian date embedded in a cherty (siliceous) rock matrix. These fossil plants undoubtedly give us the best picture yet obtained of an early type of land flora. Indeed, there are many technical features in the structures of the Rhynie plants which make it reasonable to suppose that they do, in certain respects, represent a hitherto "missing link" between a water flora and a land flora. The discovery of this Devonian bog flora is so important that a brief description of its more important plants is well worth giving. Four distinct species of vascular plants (i.e. plants with woody elements for conducting sap) have been found in the Rhynie chert. Two have been described under the generic name *Rhynia* (*R. gwynne-vaughanii* and *R. major*) differing from one another mainly in size, proportions, and a few other characters. They are small plants, with slender, leafless stems, forked branching, and reproductive organs (sporangia) developed at the ends of the branches. Near the stem-base numerous hemispherical protuberances are developed in *R. gwynne-vaughanii* and may represent primitive leaves. The stems have a well differentiated anatomical structure with wood, bast, and outer layers of tissue. Over the outer skin there is a well-marked thickening (cuticle) which is definitely a character associated with life on the land. The spores have shape and markings recalling those of the fern group. In *Hornea lignieri* there is an underground tuberous structure without conducting strands, which, however, are present in the erect smooth stems. The spore cases are particularly interesting since they possess a column of sterile tissue running up from the base into the cavity—a moss-like character. *Asteroxylon mackiei* is more complicated than the other Rhynie plants. In general appearance, and in some anatomical details, it resembles a club-moss. There is a smooth underground stem or rhizome while the branched aerial axis is closely covered with very narrow and spirally arranged leaves, though the spore cases were probably borne on naked stalks. The stems have a central mass of wood which is star-shaped in cross section ; hence the name of the genus, *Asteroxylon.*

There is no doubt that these Rhynie chert plants are of relatively very simple structure for land plants and there seems no reason for doubting that this simplicity is primitive and not due to reduction. Further, the genera mentioned, together with others placed now in the same group, but from different localities and of different ages, show combinations of characters which in later floras are separated and limited to distinct groups such as those of the fern allies and mosses. The group as a whole may well prove to contain the 250-million-year-old ancestors of our existing land plants.

In the Upper Devonian period true ferns and fern allies existed and quite probably also early types of seed-bearing plants with the habits of ferns. The material, most often imperfectly preserved, suggests a gradual evolution to the rich, varied and well-investigated flora of the next period.

CARBONIFEROUS

The Carboniferous is, in the British Isles, the period when our coal beds were deposited. As is well known, our ordinary coal consists of the carbonized remains of a dense vegetation growing in swampy lowlands. Our knowledge of the plant life of Coal Measure times is relatively great, at least so far as concerns the plants which grew in or near what are now our coal basins. It is, however, usually impossible to distinguish actual plant parts in a lump of household coal since the changes the tissues have undergone are too great. It is from the shale beds associated with the layers of coal that plant fossils are obtained in great numbers, as impressions, casts, or embedded in a calcified condition in "coal-balls." By cutting thin sections through these coal-balls, and mounting them in Canada balsam on glass slides, excellent microscopic preparations can be made. These show many details of anatomical structure both of vegetative and reproductive parts. A more modern, quicker, and cheaper method is the use of collodion films which after taking an impression from a smoothly cut surface can be peeled off. We owe our considerable knowledge of Carboniferous plant life mainly to the abundance of plant fossils and their varied and often excellent preservation, but the economic value of coal beds has also been a factor in aiding research since the fossil flora is of use in recognizing and tracing the most important seams.

Some of the plant groups, in the major sense, composing the Coal Measure flora, are still in existence, but others are extinct. None of the species, genera, or families exists now anywhere in the known flora of the world. Both woody and herbaceous types are found. The vegetation was extremely lush and the characters of the plants suggest a mild to subtropical, humid, and above all equable climate.

The lycopods of to-day consist of herbaceous plants of which a few species of *Lycopodium* and *Selaginella* occur in our own British flora. In the Carboniferous period, however, they were a dominant group in the flora both in numbers and in the size to which many of them grew. Thus, species of *Lepidodendron* and *Sigillaria* grew as tall much-branched trees, with thick trunks attaining heights of 100 feet or more. The leaves were numerous and narrow, though sometimes long, in some species even attaining a length of three feet. Reproduction was by means of spores (not by seeds) and the spore-cases were borne at the base of special leaves (sporophylls) generally aggregated into cones. Most of the genera produced two kinds of spores in separate spore-cases—often, however, in the same cone. The smaller spores, usually produced in enormous numbers, gave rise to the male organs and the larger ones to the female organs. These fossil lycopods undoubtedly contributed greatly to what are now seams of coal, and in some parts of the world whole layers of coal are composed solely of their carbonized spores.

The horsetail group is represented in our living flora only by the herbaceous genus of the horsetails proper, *Equisetum*. In Carboniferous times there existed the plants placed in the genus *Calamites* (in the broad sense[1]) which showed some of the characters still retained by our horsetails, such as the arrangement of the leaves in whorls. The Carboniferous representatives of the group, however, developed masses of wood in the stems to the extent that their habit was tree-like although the general form was very different from that of the trees of our existing British woodlands. With a few exceptions, the leaves were simple and

[1] " In the broad sense " indicates that the genus, or other systematic group, is sometimes, or probably can be, split up into a number of smaller genera or other groups.

more or less strap-shaped. All members of the group reproduced by spores and not by seeds, though some had one and some two kinds of spores.

Another group, in some respects allied to the horsetails, but now quite extinct, is represented by the genus *Sphenophyllum*. A number of species are known and appear to have been herbaceous creeping plants with slender jointed stems on which were borne whorls of wedge-shaped leaves, usually six in number at every node. The cones show some differences in structure according to the species but the spores were all of one kind. A very remarkable cone, named *Cheirostrobus*, was found at Pettycur on the Firth of Forth, in the Calciferous Sandstone series at the base of the Carboniferous. The cone is well preserved and its structure has been carefully worked out, though nothing is known as to the vegetative structure of the plant which produced it. The special interest of this cone is that though it is one of the oldest fructifications whose structure is adequately known (at least apart from the fossils of the Rhynie chert referred to above) it is perhaps the most complex of all cones so far found. It has been called a " synthetic " type, in the sense that it shows a combination of characters which are distinctive for a number of groups of plants.

There is no doubt that true ferns occurred in Carboniferous times ; but though impressions of " fern-like " foliage are amongst the commonest of fossils in rocks of the Coal Measures, we now know that many of them were not ferns. Two of the oldest types of true ferns are represented in living floras by the tropical ferns called *Marattia* and allied genera and by the fern-royal (*Osmunda*). The groups (in the broad sense) to which both of these belong have left fossils of Carboniferous age. The description of Coal Measure times as " the age of ferns " is now, however, proved to be incorrect, for though ferns occurred they were not a dominant feature of the vegetation. Many of the " fern-like " leaves already mentioned were borne on plants of fern-like appearance and vegetative structure but which reproduced by seeds or seed-like structures and not by spores. The seed-habit is one definitely suited to life on land and could only have been evolved after a land vegetation had become established. For instance, it involves, apart from a few very special and derived examples, pollination by means of pollen-grains dispersed through the air and germination of the seeds in soil. The evidence points strongly to the evolution of the seed-habit from spore-bearing plants with woody elements in their tissues (technically termed the vascular cryptogams or

Pteridophyta) which were either ferns or an extinct group with many fern characters. Another point of considerable evolutionary interest is that though the seeds of the "seed-ferns," or Pteridosperms, show some characters we have to consider relatively primitive, their structure is not simple. Rather it is highly complex and in many ways more complex than that of existing seed-plants. The "seeds," so far studied, of these seed-ferns are unfortunately all immature in that they do not contain embryos. The majority of these "seeds," or perhaps it would be better to term them "ovules," i.e. structures that become seeds after fertilization and the development of an embryo, were detached from the plant; but it is uncertain whether this was an "accident" of common occurrence, in some ways equivalent to the fall of immature fruits in apples, plums, and other fruit trees, or whether development of embryos was normally delayed till after the shedding of the ovules. Whether or not these seed-ferns were ancestral to other groups of seed-bearing plants and ultimately to our existing flowering plants, the fact remains that they were the first plants we know to produce true ovules and true pollen.

Another group of Carboniferous plants consists of trees which definitely reproduced by seeds. The group-name Cordaitales is usually applied to it, and it is now extinct. Some of the members appear, in several respects, intermediate between the seed-ferns and the group of seed plants known to botanists as the Gymnospermae, which includes such familiar plants as pines, cedars, and junipers, as well as the less familiar cycads. Other, and these the more typical, Cordaitales, are quite definitely Gymnospermae (see p. 263 for a diagnosis of this major class). The name-genus, *Cordaites*, had tall, slender trunks and long, narrow leaves with parallel veins giving a habit very different from that of any existing British trees. The reproductive bodies were compacted into cones composed of bracts subtending stamens, in the male cones, or naked ovules, in the female cones.

In the above account of the Carboniferous flora, only some of the more important and best-known plants or plant-groups have been very briefly described. Enough has, however, been said to prove that in this far-off period there had developed a very definite and varied land flora and that this included not only representatives of both extinct and still-existing groups of spore-bearing plants but also seed-bearing plants.

Permian

The close of the Carboniferous period was marked by extensive movements of the earth's crust both in this country and in much of Europe. The folding gave rise to a great mountain system, which, amongst other results, led to the separation of our coal basins. On the crumpled land surface, after probably a considerable interval of time, Permian sandstones, conglomerates, and magnesian limestones were laid down, but these occupy a relatively small area of surface in the British Isles and are here poor in plant remains. What fossils have been found can be regarded as those of types surviving from Carboniferous times.

Triassic

Rocks of Triassic age in the British Isles occupy a considerable area in the west and north-west midlands of England. They appear to have been laid down under conditions of almost desert-like dryness and only here and there are plant fossils found in any abundance. The club-moss, horsetail, and fern-like groups of the Coal Measures still just survived but there is evidence of a considerable increase in importance of Gymnospermae, both coniferous (like *Voltzia*) and cycadaceous (as *Pterophyllum*). A good deal more is known about Triassic fossil plants from rocks of this age in other countries, notably from Greenland and the southern hemisphere, but these are outside the scope of this book.

Jurassic

There is a broad belt of Jurassic rocks running across the centre of England from the coast of Dorset in a north-easterly direction to east Yorkshire. The two main divisions are known as the Lias and the Oolite, the former being the older. While club-mosses, horsetails, and ferns existed in considerable diversity there is no doubt that the dominant types of plants were Gymnospermae. These were of numerous and varied kinds, some more or less allied to existing types of cycads (*Cycas*, *Zamia*, *Encephalartos*, etc., such as can be seen in the Palm House at Kew), others possibly ancestral to existing genera of conifers, and still others (such as the important group known as the Bennettitales) now quite extinct. An important structural character of some of these last is that they possessed " bisexual flowers," or, more simply and possibly more accurately, reproductive branches in which stamens (in one sense male organs, in that they give rise to pollen-grains with male sex cells inside them) and ovule-bearing organs (in one sense female, in that a female sex cell is found in every ovule) were so associated that the former arose on the axis below the latter ; to this extent they agreed with modern " two-sexed " flowers. It may be noted that the well-known " fossil crows'-nests " to be seen, for example, on some of the Dorset cliffs, as just east of Lulworth Cove, are the petrified remains of stems of plants of general cycad affinity.

It was thought at one time that a group of plants, known as the Caytoniales, described from Jurassic beds in Yorkshire, had ovules completely closed before pollination, but it now appears that they had direct pollination of ovules. They represent, in the genera *Caytonia* and *Gristhorpia*, a very distinct and now extinct group of plants whose relationships remain doubtful. These plants had ovules more or less surrounded by cases which, if not completely closed, at least approximated, as protective organs, to ovaries.

It is very probable that true flowering plants (Angiospermae) existed in Jurassic times but undoubted fossils of this group are first found in the next period. It may be that the earliest flowering plants lived in localities and under habitat conditions such that their remains could not be fossilized, for example, on mountains away from swamps and the sea coast. Nevertheless intensive research on Jurassic deposits may yet reveal their presence and help to settle many controversial puzzles, some of which are discussed later in this chapter.

Summary and Conclusions Regarding Early Floras

There is strong evidence that the earliest plants were such as could only exist submerged in water ; whether fresh, brackish, or salt is a question of less importance, although it may be noted as probable that sea water was less salt in the early Palaeozoic than it is now. From aquatic ancestors the land flora arose. The earliest " land " plants were, at least in part, swamp or bog plants and the flora of the Rhynie chert provides us with one set of hitherto " missing links." The land habit necessitated inside conducting strands along which water and food materials in solution could travel since these could not be absorbed directly by parts growing up into the air. Hence the rise of " vascular " plants or those with wood for conducting watery solutions and bast for conducting food made by the plant. They still, however, reproduced by spores, and liquid water was necessary for fertilization.

The second great advance shown in Palaeozoic floras was the evolution of the seed habit. The most primitive seed-bearing plants were the "seed-ferns" or Pteridospermae in which there was a combination of vegetative characters recalling those of ferns and fern-like plants (vascular cryptogams or Pteridophyta) with reproduction by means of seeds. The structure of the ovules of these very early plants is complicated and shows clearly that liquid water was still essential for fertilization. Either from Pteridospermae or from some other and perhaps still unknown ancestor there evolved the true Gymnospermae: the now extinct Cordaitales and groups allied, if more or less remotely, to the still existing cycads and conifers. The Gymnospermae became increasingly dominant in the early and middle parts of the Mesozoic and were then represented by a wealth of genera and species. It is interesting to note that in this group the last structural feature linking back to the far-off aquatic origin of plant life is discarded. Certainly a considerable number of Mesozoic Gymnospermae required, and in the existing flora of the world the cycads and maiden-hair tree (*Ginkgo*) still require, liquid water for fertilization. It is only in the true conifers, as represented by pines, junipers and the yew (Pl. 1, p. 34) and the Angiospermae that there are non-motile male sex cells which reach the ovule entirely by means of a pollen-tube. The following scheme shows the main facts of this change:

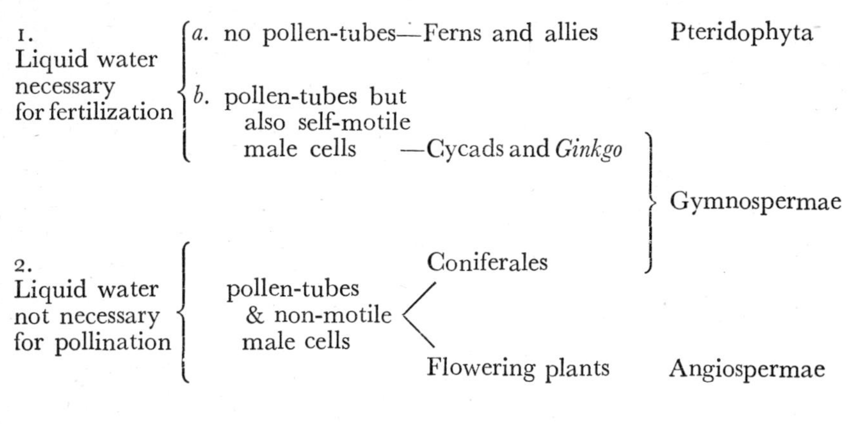

Flowering Plants

The transition from floras dominated by Gymnospermae to those dominated by Angiospermae would appear, from the existing fossil evidence, to have been sudden, but this evidence is meagre. In most of the earlier Cretaceous rocks, as, for example, our own British Wealden, there is no evidence of Angiospermae, but Jurassic types, and even species, of ferns, cycads, and conifers continued to flourish. The oldest fossil evidence at present known for Angiospermae is in the form of pollen from Jurassic rocks determined as belonging to Nymphaeaceae, Magnoliaceae, and (?) Juglandaceae. Angiosperm fossils, some problematical or doubtful, have been recorded from Lower Cretaceous rocks in North America and Greenland. It is in rocks of Lower Greensand (or Aptian) age that we have in Britain the first definite proof of "flowering plants." Petrified wood of five different genera, belonging to various families of Dicotyledones, have been described. An interesting feature is that the anatomical structure of these fossil woods is surprisingly modern and leaves no doubt as to the antiquity of angiosperm characters, at least so far as structure of wood is concerned; and botanists have every reason to lay considerable stress on this since it is regularly correlated with important classificatory structures of other vegetative organs and with reproductive parts. It is most unlikely that these Lower Greensand fossils represent the most primitive flowering plants. The place or places and time or times of origin of these remain unknown. Suggestions as to their ancestry have been fairly numerous. The one that has probably received the widest, though often tentative, support is that they arose from cycad-like (Bennettitalean) plants and that the most primitive of existing flowering plants are members of the magnolia family. There are, however, difficulties in accepting this view, especially as regards the structure of the female part (technically called the gynoecium) of the flower. It can, however, be accepted that the magnolia family is relatively an ancient one, though fossils representing other and quite different families are as old or older, so far as the imperfect fossil evidence at present indicates. Thus the plane (*Platanus*) and walnut (*Juglans*) go back at least to Lower Cretaceous times.

From Lower Greensand times onwards land floras were dominated by flowering plants wherever conditions were such that rooted plants could exist. In Britain, however, the younger Cretaceous deposits (as the chalk) were not laid down under conditions favouring the preservation of land plants as fossils and we proceed, therefore, to the Cainozoic or Tertiary epoch.

The lowest stages of the Tertiary may be considered together. Eocene beds, especially the London Clay of the London basin and sands and clays of the Hampshire coast (from Bournemouth to east of Barton) have yielded great numbers of fossil plants, especially fruits and seeds. Many of these have been very carefully determined and indicate that the flora of these times, in what is now south-central and south-east England, was composed of flowering plants mostly belonging to families still existing. The floristic relationships, however, are not with our present-day British flora but with that of Malaya. This suggests very strongly that conditions were then much warmer than they are to-day, indeed that they were more or less tropical. A large proportion of the fossils represent woody plants. This may be partly due to the smaller chance of preservation of the more delicately formed herbs though it must also be remembered that tropical vegetation often has a higher proportion of woody to herbaceous species than has temperate vegetation. It is possible that herbs have usually been derived, in the course of evolution, and within the flowering plants, from woody ancestors, but the evidence for this is mainly indirect and other views are held. Upper Eocene beds, as those of the Hordle cliffs of the Hampshire coast, show Eastern Asiatic and American floristic relationships. Oligocene rocks with abundant plant fossils occur in the Isle of Wight (Bembridge Beds) and at Bovey Tracey (Devon). The plant assemblages again suggest relationships with the existing floras of East Asia and America. In both the Upper Eocene and Oligocene floras some genera now typical of the European flora are found. The following table gives a few of the more important or interesting genera of British Eocene and Oligocene fossil plants :

LONDON CLAY (Lower Eocene) : *Sabal* and *Nipa* (palms) ; *Magnolia* ; *Cinnamomum*, *Litsea*, and other members of the laurel family ; *Vitis* spp. (vines) ; *Mastixia* (dogwood family) ; and *Dracontomelon*, *Spondias*, and other members of the Anacardiaceae or cashew-nut family.

BOURNEMOUTH (Middle Eocene) : *Araucarites* (allied to the monkey-puzzle tree) ; *Sequoia* (the genus of the redwoods and giant trees of California) ; *Taxodium* (the genus of the deciduous swamp-cypress) ; genera of the

Lauraceae (laurel family) as *Aniba*, *Neolitsea*, *Lindera*, and *Cinnamomum*; *Rhodomyrtus* and *Tristania* (myrtle family); and *Nothofagus* (the "beech" of the Southern Hemisphere).

HORDLE CLIFFS (Upper Eocene): *Salvinia* (a water fern); *Sequoia* (conifer); *Nipa* (a palm); *Rubus* (a blackberry); *Diospyros* (an ebony); and *Brasenia* (a water-lily).

BOVEY TRACEY (Oligocene): *Magnolia*; *Vitis* spp. (vines); *Nyssa* spp. (palms); *Taxodium*, *Sequoia*, *Taxus* (yew); *Potamogeton* (pond-weed).

BEMBRIDGE BEDS (Middle Oligocene): *Sabal* (a palm); *Ficus* (a fig); *Ranunculus* (a buttercup); *Cinnamomum* (a cinnamon); *Aldrovanda* (sundew family); *Catalpa* (an Indian bean); and *Abelia* (honeysuckle family).

John Markham

YEW, *Taxus baccata* (Taxaceae). Little Berkhampstead, Hertfordshire. October 1946

Brian Perkins

a. GRAPE HYACINTH, *Muscari racemosum* (Liliaceae)
Near Cherry Hinton, Cambridgeshire. April 1946

Eric Hosking

b. SNOWDROPS, *Galanthus nivalis* (Amaryllidaceae)
Near Woodbridge, Suffolk. February 1945

CHAPTER 4

THE ICE AND ITS INFLUENCE ON THE BRITISH FLORA

THERE IS in British deposits a considerable gap between the older Tertiary, as represented by the Oligocene rocks of the Hampshire basin, and the younger Tertiary as represented by Pliocene deposits in East Anglia. Thus, in the British Isles there are no important strata which are fully accepted as of Miocene age. One can, of course, fill in the gap from studies made on the numerous outcrops of rocks of this age on the European mainland, many of which contain determinable plant remains. It is of interest to compare the plants recorded from Eocene and Oligocene deposits with those from British Pliocene deposits. We may note that, in general, the early Tertiary plants are placed in genera which still exist, though not always in Europe, or in genera apparently closely related to such, while the species themselves are now extinct. In drifts, now accepted as of middle Pliocene age, filling fissures on the Durham coast at Castle Eden, there were found remains of 114 species of plants, mostly as seeds. Of these, 58 could be related to living allies and of these about 64 per cent were exotics, and about one-third suggested relationships with Chinese or North American types. The well-known Cromer Forest-Bed outcrops here and there near the foot of the cliffs between Mundesley and Sheringham and from it excellently preserved plant-remains can easily be obtained. It has been carefully investigated and the majority of the plants are not only those of existing species but of species occurring in our present British flora. In fact only 5 per cent exotic and extinct species have been recorded. The Cromer Forest-Bed is assigned to the uppermost part of the Pliocene and its fossil plants to a certain extent represent the flora of East Anglia immediately prior to the Ice Age. It does not follow that all the species represented in the Cromerian beds and now existing in Britain necessarily survived the Ice Age in our country, since they may have been locally exterminated and later re-introduced. The plant remains were probably deposited as drifted trunks, branches, foliage, fruits, and seeds washed down into what was a delta of a greater Rhine. Amongst plants not now growing as natives in Britain were the spruce (*Picea abies*) and the water chestnut (*Trapa natans*). A large number of the species were aquatics and marsh plants and these included pond-weeds (*Potamogeton*), sedges (*Carex*), water lilies (*Nuphar* and *Nymphaea*), and mare's-tail (*Hippuris vulgaris*). There were also many trees and shrubs, amongst which may be mentioned Scots pine, yew, hornbeam, hazel, birch, alder, beech, oak, elm, mountain ash, sloe, hawthorns, maple, alder buckthorn, and dogwood.

In the same Norfolk cliffs but at a higher level than the Forest-Bed is an Arctic plant-bed which shows the oncoming of the Ice Age and is regarded as the base, in this country, of the Quaternary Epoch. We thus reach the most recent of the great geological happenings which have influenced British plant life. There is still much in dispute regarding the Ice Age and its effects on plants and plant distribution. It is particularly unfortunate that there is wide divergence of opinion on matters which are in the main beyond the scope of the biologist but which have important bearings on biological problems. Thus the number of glacial and interglacial periods, the areas covered by ice at any one glacial period, and the nature of the climate during the glacial and interglacial periods are all uncertain. The botanical evidence itself is relatively meagre and has been differently interpreted.

Briefly, the position is that there are two extreme schools of thought. One holds that glaciation was so extensive, and climatic conditions during the glacial periods so unsuitable to plant life, that practically all the pre-glacial vegetation was destroyed and only a few arctic-alpine species of the flora survived. The other school maintains that conditions, at least around the margins of what ice-sheets there were at any one period and on more or less numerous ice-free areas, were not severe enough to exterminate more than a few species in Britain and that a majority, or at least a very large number, survived. To give and critically to examine all the arguments would more than fill this volume. There is still much research to be done in connection with this problem, and until an intensive study has been made of the ecology, genetics, distribution, and relationships of all the individual species concerned any generalized statement must be tentative. Even closely related species have had different histories in regard to the Ice Age. Thus, the conclusion is reached from the close study of a great mass of data that of the two bladder campions, *Silene maritima* (Pl. 38b, p. 179) survived the Ice Age in this country while *S. cucubalus* (Pl. 38a) has immigrated or been introduced since.

The following seem the best conclusions which can be accepted as working hypotheses at present. During the latter part of the Tertiary the climate of the British Isles became gradually less sub-tropical and more cool temperate until in the period represented by the Cromer Forest-Bed it was very much like that of to-day. Correspondingly, the flora was greatly though gradually changed from one with many sub-tropical or warm temperate elements to one not greatly dissimilar to that which is still found in some parts of Great Britain. It does, indeed, seem probable that the composition of our existing flora was more fundamentally influenced by the climatic changes which preceded the Ice Age than by those of either the Ice Age itself or of post-glacial times. This does not, of course, mean that the flora as a whole necessarily survived where it now occurs. This was certainly impossible over large areas for long periods. Hardier species may have survived on areas which remained free from ice and these may not have been the same through the whole Ice Age or in all the glacial periods in so far as these can be distinctly recognized. Such ice-free "islands" surrounded by land ice are often termed "nunataks." Probably more important was survival of a proportion of heath and moor herbs of our flora in the western and southern parts of Great Britain. Many of the Scottish islands, for example, were unglaciated, as also was England south of the Thames. How severe was the climate in these parts during the glacial periods, and, for that matter, during the interglacial periods too, is uncertain. It seems most likely that the greater part of our tree and much of our shrub vegetation and most of our species with a southern range did not survive the Ice Age in this country. This must mean that during the Ice Age a large proportion of our flora existed in unglaciated parts of Europe, most probably near the Atlantic, and migrated in post-glacial times across what is now the English Channel but was land till after the Ice Age.

Only a brief reference will be made here to a few special groups of plants since the range of some of them and of others will be discussed later in this work. The age, in this country, of the so-called Lusitanian plants, which are limited mainly to Ireland and south-western England in the British Isles, but extend, partly discontinuously, to Spain and Portugal and even farther into the Mediterranean Region, is debatable. Most probably the group is heterogeneous in relation to its history in the British Isles and every species requires detailed investigation of a much more intensive nature than it has yet received.

It is difficult to conceive of the strawberry tree surviving the Ice Age even in south-western Ireland, while some of the saxifrages are probably much more hardy than the strawberry tree. As regards our endemic species, and those whose range centres in Great Britain, their history is certainly not always similar and to group them together in any other sense than as more or less limited in their present range to the British Isles is misleading. Some have certainly originated since the Ice Age ; a few may be survivals from the pre-glacial flora ; some probably originated during the Ice Age ; some are possibly immigrants from the Continent.

The influence of the Ice Age, and still more of its oncoming in Pliocene times, on the British vegetation and flora, was profound, but the question of survival in the British Isles is of limited and secondary importance because the whole flora of northern and central Europe was affected and Great Britain was then land-connected with the Continent. To attempt a full investigation of all the problems would involve wide considerations of the European flora as a whole and would also take us out of the realm of plant life into zoology, geology, meteorology, and other subjects.

CHAPTER 5

POST-GLACIAL FLORISTIC CHANGES

A VERY great deal of research has been devoted in recent times to a study of post-glacial vegetation, or, in other words, of the natural plant covering between the end of the Ice Age and the present day. In this country, as in many other parts of northern and central Europe, there are peat bogs which have been formed through a longer or shorter period of post-glacial time. In these bogs, plant remains form the great bulk of the material and very often the actual species of plants which grew on the bogs or in woods in their neighbourhood can be determined from their remains. Roots, rhizomes (underground elongated stems), stems, leaves, fruits, and seeds may be specifically determinable. A great many bogs are built up largely of mosses, especially of species of bogmoss (*Sphagnum* spp.), and these too are sometimes determinable from their semi-fossilized remains in the peat. Extremely interesting results have been obtained from a study of pollen-grains preserved in the bog deposits. The walls of pollen-grains are almost indestructible by such natural processes of decay as occur in peat bogs. The contents are destroyed, but the size, shape, and markings of the outer coat give reliable and easily recognizable characters by which the genus and sometimes the species can be determined. Pollen-grains are often produced and preserved in enormous quantities, and one great advantage in utilizing them, in the reconstruction of the vegetation which occurred in a given area when the peat of a particular layer or horizon was forming, is that quantitative estimates of abundance are possible. By taking suitable precautions in the collecting and analysing of samples the percentage of pollen-grains of different species (or genera) may be accepted as giving a picture of the relative abundance of the species (or genera) present. As will be explained in detail in a later chapter, forests have a special importance in vegetational studies because they are essentially composed of trees which by their height, size, and spread of canopy dominate the habitat more effectively than do the shrubs or herbs associated with them. In fact, trees largely make the peculiarities of the habitat for other woodland plants by cutting off much of the light, providing a humid atmosphere, breaking the wind, producing materials for layers of leaf-mould, and so on. For most of Britain, forest is the natural " climax " and during a great part of post-glacial time forests covered most of the surface till man cleared them away. Further, we have few native forest trees and most of them (as Scots pine, birch, and oak) are wind-pollinated and produce vast quantities of pollen-grains. It is thus reasonable that research has so far been mainly concentrated on a study of tree pollen in peat. A large number of peat deposits have now been examined for tree pollen in England and Wales and some in Ireland. Moreover, it has now been found possible to correlate the results and to present a general sequence of the changes in the tree vegetation over much of the southern part of Britain since the Ice Age.

A simplification of the latest zonation of British peat deposits is given below, starting from immediately after the Ice Age or even before the ice had finally melted.

1. *Lower Dryas clays.*
2. *Late glacial birch period.*
3. *Upper Dryas clays.*

These three zones have been described from Denmark and from Ireland and are probably represented by equivalents in some British bogs. Zones 1 and 3, as judged from their plant remains, were formed under arctic conditions while zone 2 represents a warmer interlude.

4. *Birch-pine zone.* In this zone there is dominance of birch pollen with pine as the other most important tree. Willow and hazel pollen-grains sometimes occur. The ratio of non-tree pollen to tree pollen is so high as to suggest a relatively open landscape.

5. *Pine zone.* This zone, with widespread replacement of birch by pine, indicates the extension of forest. Hazel increases rapidly and occasionally warmth-needing trees are represented.

6. *Pine-hazel zone.* In general, pine and hazel are dominant but oak and elm occur and finally lime and alder show a tendency to extend.

7. *Alder-oak-elm-lime zone.* Pine is suddenly replaced by alder as the most abundant tree pollen. Oak maintains a fairly high value or even increases. Elm decreases about half-way through the zone. Lime expands considerably. Small amounts of beech pollen occur sporadically.

8. *Alder-oak-elm-birch (beech) zone.* Birch pollen becomes more abundant while lime either ceases or becomes discontinuous. Beech (and sometimes hornbeam) is present in much greater amount than formerly.

This classification is a generalized one and modifications of it due to local conditions are known. Further, there is evidence of a differentiation of the country into a southern and eastern portion (roughly south-east of a line from Somerset to north Yorkshire) and a western and northern region. Birch tends to increase northwards and hazel is most abundant in the west. There is in this distinction a broad agreement with still recognizable climatic and vegetational regions best designated as south-eastern, western, and northern. It would appear that such differentiation into vegetational regions is ancient and relatively permanent. Climatic changes have induced equivalent, though not identical, changes in forest composition in the different parts of England and Wales since the Ice Age.

Another matter upon which the pollen analysis of bogs has thrown light is the native status of our trees. Many trees, such as the spruce (*Picea*), fir (*Abies*), and horse chestnut (*Aesculus*), now common in British woodlands, are known to be introductions. Others, such as the birches, alder, and oaks, have been generally accepted as native. There has, however, been much controversy over the native status of beech, elms, limes, and (in England) Scots pine. It now seems well established that beech is native in England and Wales as is also one or more species of elm. Of the limes, only one species, *Tilia cordata*, has been identified by pollen-grains from peat. Scots pine too appears to be truly native in England and Wales. There is little doubt that pine trees now growing on some old bog surfaces are the direct descendants of native trees. This, of course, does not mean that every pine wood is natural and native, still less primitive, and the history of every one needs separate consideration. Unfortunately it has not yet been possible to separate the pollen-grains, in British bogs, of the white birch (*Betula pendula*) from those of the hairy birch (*B. pubescens*), or those of the common or stalked oak (*Quercus robur*) from those of the sessile oak (*Q. petraea*). Also it is not possible at present to say what species of elm are represented by the elm pollen of bogs.

Though pollen analysis of peat bogs is the most recent, and in many ways a very satisfactory, method of investigating the post-glacial history of our vegetation, it is not the only method. There is no doubt it will be extended in the future to include the pollen of plants other than dominant trees and a few shrubs. There are likely also to be advances in the earlier-used methods of peat analysis in which plant remains other than pollen-grains were determined. Throughout studies of post-glacial vegetation attempts have been made to correlate beds of plant remains with archaeological material, which can be more or less dated, and also with climatic changes. A widely used scheme of post-glacial climatic periods, originally evolved for Scandinavia, has been applied frequently in different parts of Europe. The following simple table attempts a correlation of various data but can only be considered as an approximate guide to subdivisions based on different

Approximate dates	*Climatic Periods*	*Pollen Zones*	*Climax Vegetation*	*Archaeological Periods*
B.C.				
9000	Sub-arctic (cold and dry)	IV	Tundra Birch with some pine	End of Palaeolithic
8000	Pre-boreal (fluctuations of climate)	V	Pine	
7000	Boreal (warm & dry)	VI	Pine & hazel with broad-leaved trees increasing	Mesolithic
6000				
5000				
4000	Atlantic (warm & wet)	VII	Alder-oak-elm-lime	
3000				
2000	Sub-boreal (drier)			Neolithic Bronze Age
1000				
—	Sub-Atlantic (cool and wet)	VII-VIII VIII	Alder-oak-elm-birch with some beech	Iron Age Historic Period
A.D.				
1000	Recent (warmer and drier)			
2000				

PLATE 3

John Markham

BLACK BRYONY, *Tamus communis* (Dioscoreaceae). Shenley, Hertfordshire. September 1945

PLATE 4

Brian Perkins

PASQUE-FLOWER, *Anemone pulsatilla* (Ranunculaceae). Devil's Ditch, Cambridgeshire. April 1946

data, and further investigations and discoveries may lead to its having to be considerably modified. In particular, evidence is often lacking in Britain of a major climatic transition from Atlantic to Sub-boreal climatic periods, though part of the Bronze Age was probably relatively dry in this country.

Evidence for the post-glacial history of the British flora, other than that derived from a study of tree and shrub pollen in peat, is meagre and what facts are known are not always given the same interpretation. There is little doubt that S.E. England was joined to the Continent for some time (say one to two millennia) after the final retreat of the ice. The country of the Straits of Dover, it has been stated, began to subside about 8,000 years ago and the formation of a sea passage from the North Sea to the English Channel was comparatively quickly completed, though it was for some time very narrow. The whole process took a few thousand years but there may have been a considerable interval, after the first break-through, before submergence was resumed and the shores were eaten back approximately to their present position.

Narrow as are the Straits of Dover, they and the English Channel have retarded if not permanently reduced the natural flow of Continental migrants. It is a somewhat similar story for the late history of the Irish flora. The Irish Sea was still much restricted in Boreal times and the subsidence of the " bridge " took place finally in Atlantic times and was thus approximately coincident with the complete formation of the English Channel. Again, the subsidence impeded western migration. Only 67·5 per cent of the species of plants of Great Britain inhabit Ireland at the present day, though different ecological conditions probably account for many of the absentees. Evidence of submergence is found in the submerged forests which occur at various places round the British coasts and are locally termed " Noah's Woods " from a popular explanation that they were connected with Noah's deluge. Oak was the principal tree. Their positions, between extreme tide-marks, suggest that, in the main, they represent the last general phases of land submergence whose earlier phases gave our islands their present general configuration. There is, however, evidence that locally, as at Romney Marsh, much more recent subsidence has occurred, of post-Roman and even post-Norman date. Indeed, in places subsidence may still be proceeding, as on parts of the Dorset coast.

We must now consider the changes in the British flora due to man's activities. Man is supposed to have entered Britain on the heels of a temporarily retreating ice-sheet in Upper Palaeolithic (Aurignacian) times, possibly as much as 20,000 years ago. In all probability Palaeolithic and Mesolithic men existed solely by hunting, fishing, and food-gathering and their numbers and behaviour were such as to have little influence on the flora and vegetation. Somewhere about 2400 B.C. there was a considerable immigration with a new economy, that of food-producing by stock-raising and tillage. By this time the British vegetation had recovered from the devastations of the Ice Age, though the flora remained impoverished relative to what it was before the on-coming cold exterminated a very high proportion of species. There is little doubt that the greater part of Britain, at least that below 400 m. and the extreme north, was clothed with forest, often dense and very largely of oak. The question of forest on the chalk downlands of southern England has been much debated. The ecological evidence, on the whole, favours the view that the grassy slopes, low hills, and undulating plains of the Downs are not the natural climax type of vegetation but owe their origin and maintenance to man and his domesticated animals. It was the chalk and limestone hills, in so far as they were not capped by clay, that were especially chosen by man for early settlement in Britain. The chalk yielded flint, the soils were light, porous and dry, the habitat conditions healthy, and the forest, often dominated by ash, on the whole more open, lighter, and easier to conquer than that in the damp lowlands on heavy soil. The lowlands, where not unhealthy, untraversable marshes, fens, or bogs, were clothed with thick forests of great oaks impossible to cut down with stone implements and difficult to destroy by fire. Such, at least, is the picture painted with reasonable imagination. Early man appears to have disliked the damp oak woods (see p. 109) even to the beginning of the Christian era. The chief domestic animals were present in Britain not later than the close of the Neolithic phase. Pastoralism combined with agriculture necessitated extensive and relatively open areas. Chalk, limestones, and certain sandy heathlands gave sufficient unbroken extents in the southern and midland parts of England. It has been possible, in Denmark, to correlate changes in the pollen content of bogs with the invasion of the country by Neolithic man and further changes with the beginning of the Iron Age. A study of general pollen content, not of tree and shrub pollen alone, of British bogs may throw much light on the introduction of cultivated crops and their associated weeds.

Our present knowledge of crops in pre-historic Britain is meagre. A grain, probably wheat, was found on a definite Neolithic site at Rothesay (Buteshire), associated with contemporary pottery and "saddle querns." Neolithic saddle querns for corn grinding are recorded from various Neolithic settlements in Britain. Spelt and emmer (*Triticum dicoccum*) have been found in Bronze Age barrows and the Late Bronze Age settlement on Mildenhall Fen (Suffolk) in the form of impressions of grain and spikelets. The closely related small spelt (*T. monococcum*) is also represented. Common wheat (*T. vulgare*) seems to have been rare in early England but has been recorded from an Early Iron Age settlement at Abingdon (Berkshire). The six-rowed barley (*Hordeum polystichum*) was the chief grain of the Cambridge area. Naked and husked grains occur with equal frequency in the Bronze Age, but the husked form was almost completely dominant in Anglo-Saxon times. Oats (*Avena sativa*) is recorded from Fifield Bavant in Wiltshire (*c.* 400–250 B.C.). There is no doubt that the practice of grain-growing had reached Britain in Neolithic times and has continued ever since. None of the cereals is native to Britain. With the introduction of crop seed there came various weeds and cultivation of the soil led to secondary successions (see p. 87) of longer or shorter duration according to the length of fallowing or the interval periods of shifting cultivation. Early introduced weeds may well have included poppies (*Papaver* spp.), Venus's comb (*Scandix pecten-veneris*), some speedwells (*Veronica* spp.), some spurges (*Euphorbia* spp.), and others. Indeed, it is most probable that a majority of our common weeds of arable land are not indigenous.

A particularly interesting find of plant remains on a Late Bronze Age hut site in Minnis Bay, Kent, throws some light on the early presence of weeds and other plants in southern England. Four general habitats are suggested by the plant remains (wood, leaves, flowers, fruits, and seeds): (1) cultivated and waste ground; (2) fresh water and fen; (3) sandy heath; (4) woodland. Species from the first habitat greatly preponderate, and include: poppy (*Papaver rhoeas*), penny-cress (*Thlaspi arvense*), thyme-leaved sandwort (*Arenaria serpyllifolia*), corn spurrey (*Spergula sativa*), purging flax (*Linum catharticum*), silver-weed (*Potentilla anserina*), burr chervil (*Anthriscus vulgaris*), coriander (*Coriandrum sativum*), ox-tongue (*Picris echioides*), smooth hawk's-beard (*Crepis capillaris*) and stinging-nettle (*Urtica dioica*). These and other species probably reflect the cultivation of crops by the Late Bronze Age inhabitants and may represent residue from threshing.

A very large number of pieces of charcoal from Neolithic, Early Iron Age, and Late Iron Age deposits from Maiden Castle, Dorset, have been examined. Species present in both Neolithic and Late Iron Age charcoals were: hazel (*Corylus avellana*), ash (*Fraxinus excelsior*), crab (*Malus pumila*), oak (*Quercus* sp.), willow (*Salix* sp.). In the Neolithic charcoals only were: poplar (*Populus* sp.), common buckthorn (*Rhamnus cathartica*), whitebeam (*Sorbus aria*). In both Neolithic and Early Iron Age, yew (*Taxus baccata*) occurred. In the Late Iron Age charcoals only were: birch (*Betula* sp.), hawthorn (*Crataegus* sp.), service-tree (*Sorbus torminalis*), elm (*Ulmus* sp.), plum (*Prunus domestica*), gean (*P. avium*), and blackthorn (*P. spinosa*). From the detailed investigation of these charcoals it was concluded that in Neolithic times the chalk of Dorset was probably clothed with a closed plant community of woodland of the oak-hazel type. Passing from the earliest to the latest period, there is evidence of some change, probably due to continued climatic action which resulted in leaching and a less calcicole flora, whilst forest destruction by man led to a more open vegetation, less favourable soil conditions, and ultimately no doubt to soil erosion. The evidence favours the view that, throughout the period represented, the climatic conditions were probably essentially similar to those obtaining at the present day. Some of these conclusions have been criticized mainly on the grounds of insufficient evidence.

The Roman occupation of much of Britain, which lasted for nearly 400 years, resulted, after a preliminary period of wars, in relative stabilization and romanization, especially of England. Population increased and agriculture was greatly extended. Forest destruction continued. The Cotswold wool industry came into or continued in existence. Iron-smelting, which began before the Roman invasion, was extended in the Weald area. There is apparently very little evidence from well-dated remains regarding plant life in Britain during Roman times. A collection of seeds from the Roman Station at Silchester, Hants, were of rather mixed habitats, common marsh plants and weeds of arable land predominating. The coming of the Saxons led especially to clearance of woods from the clays of the lowlands and valleys. The higher grounds became largely derelict and some reverted to woodlands. The

Brian Perkins

WILD CARROT, *Daucus carota* (Umbelliferae)
Gog Magog Hills, Cambridgeshire. August 1946

Robert Atkinson

GREEN HELLEBORE, *Helleborus viridis* (Ranunculaceae)
Near Nuffield, Oxfordshire. March 1946

Normans, with their passion for hunting, applied forest laws to large areas of Royal Forest, parts of which are still woodland, as Epping Forest, the Forest of Dean, Windsor Forest, and the New Forest. There was deforestation on a large scale in the latter part of the twelfth century and this continued throughout the Middle Ages. The well-known manorial system, with its division into arable land cultivated on the strip system, common or grazing land, and woodland, continued till the time of the earliest " enclosures " and locally much later. Forest destruction increased with the need for timber and for charcoal for smelting and glass-making. The dearth of timber was first felt acutely in Scotland and the north of England. Rabbits were probably introduced by the Normans but it was not until the middle of the fifteenth century that their numbers approached those of recent times, with the consequent influence on the vegetation. Sheep-raising proved most profitable in Britain through the Middle Ages down to modern times. In the middle of the fourteenth century the number of sheep is estimated to have been eight million for England alone, when the human population was two million. At the end of the seventeenth century the estimate is twelve million sheep and four and a half million cattle. In attempting to determine the influence of these flocks and herds on the vegetation it must be remembered that there were then no imported " cake " or other foreign foodstuffs, while turnips and other roots were hardly known and never cultivated as farm crops, and the same was mainly true of legumes like the clovers. The animals were grazed on pastures, common or enclosed, and in stall were fed on hay from the meadows and the foliage of trees and shrubs. Grazing animals effectively prevent rejuvenation of woodland. Drainage of marshlands reached a climax in the main drainage of the fens in the seventeenth century. At about the same time the forests of the Scottish Highland valleys and glens were ruthlessly exploited, the oaks at lower and the pine and birch at higher elevations and on poorer soils being destroyed in a wholesale manner. At the end of the eighteenth and through the nineteenth century sheep-farming spread in the hill regions largely replacing cattle-rearing and resulting in the great spread of bracken.

Hunting in the style of the Normans and, with modifications, down to the introduction of more or less modern sporting guns influenced vegetation by the reservation of large " forests " which were by no means all woodlands with tall trees. More modern shooting, with its strict game laws and the maintenance of cover for protected, conserved, and often artificially reared pheasants, resulted in woods and copses which, more the concern of the gamekeeper than of the forester, preserved a great deal of wild life. The advent of modern forestry with the planting of large areas of exotics will lead to further and not yet clearly foreseen changes in the plant life of very considerable areas.

CHAPTER 6

THE PRESENT COMPOSITION OF THE BRITISH FLORA

BY "FLORA" is meant the different kinds of plants occurring in a given area. Strictly speaking all kinds of plants together make up the flora, that is Bacteria, Fungi, Algae, lichens, mosses, liverworts, ferns, and fern-like plants should all be included as well as the seed-bearing plants. In this volume, for a variety of reasons, emphasis is given to the seed-bearing plants though references to other groups are sometimes made. Certainly the seed-bearing plants are dominant in British plant life from most points of view, at least on land. Thus they include the plants of largest size, those most conspicuous in and biotically controlling almost all the main plant communities, and those of greatest economic importance. It is probable that the cryptogams (as the plants not reproducing by seeds are termed) exceed the seed plants in numbers of individuals as they do in numbers of kinds if by "kinds" be meant taxonomic species. The following table gives round-figure estimates of the number of species occurring in the British flora of the different major groups of plants. These figures may be taken as roughly comparative but as will become apparent later the term "species" has no absolutely fixed value.

Group	Species
Bacteria	500
Fungi	7,000
Algae: Fresh-water and terrestrial	1,500
Marine (excl. plankton)	730
Lichens	1,200
Mosses	630
Liverworts	280
Ferns	50
Fern-allies	20
Seed-plants	1,500
Total	13,410

It is obvious that in order to deal with such very large numbers of different kinds of plants a system of classification of the species within each of these large groups is essential. A considerable number of class categories have been more or less widely used. The size of groups, in the sense of the number of species included, varies enormously and, in general, the larger the group the greater is the number of categories needed, although other considerations also play a part. Two classes above the species are essential with existing classifications, for most, if not all, groups of plants and, with only special exceptions, will be found sufficient for our studies of British plant life. These are the genus and family. Species showing certain characters, most often of flower, fruit, and seed, in common are grouped into one genus, and genera with certain common characteristics are grouped into one family. The interesting theoretical questions of classification must be deferred to a later chapter. Here we are concerned with some instructive results which we owe to the labour of plant systematists.

The table opposite includes only the seed-bearing plants. It has been compiled mainly from Druce's *List of British Plants*, 2nd ed. (1928), with certain emendations and modifications. The sequence of families is approximately that of Bentham and Hooker's classification and the figures may be taken as generally acceptable if some allowance be made for diversity of opinion as to where boundaries can best be drawn between some species and between some genera, and as to when a species is sufficiently established to be considered a member of the British flora. Microspecies of such genera as *Capsella*, *Hieracium*, *Taraxacum*, and *Rubus* have not been included in the counts.

Many criticisms could be made of the details of this table. Those who take a very narrow view of "species" would complain that the figures are too low. On the other hand if all weeds of cultivation and plants of waste places near habitations (ruderals) were excluded the number would be reduced to about 1,200, if the same species standard were kept. Nevertheless, a number of interesting facts emerge. Of the 91 families, 34 are represented by only one genus each. This is one indication of the impoverishment of the flora which resulted from the on-coming of the Ice Age, the more so that many of these families which have only one genus in the British flora have numerous genera in other parts of the world (as Violaceae, Rhamnaceae, Cucurbitaceae, Araliaceae, Verbenaceae, and Araceae). The largest family is the Gramineae, with Compositae second, for both genera and species. If all small species (microspecies) were included, Compositae would be the largest family in number of species and Rosaceae would be second,

	Family	Genera	Species
1	Ranunculaceae	12	45
2	Berberidaceae	1	1
3	Nymphaeaceae	2	3
4	Papaveraceae	3	7
5	Fumariaceae	3	13
6	Cruciferae	22	59
7	Resedaceae	1	2
8	Cistaceae	1	4
9	Violaceae	1	19
10	Polygalaceae	1	7
11	Frankeniaceae	1	1
12	Caryophyllaceae	12	67
13	Portulacaceae	1	2
14	Elatinaceae	1	2
15	Hypericaceae	1	12
16	Malvaceae	3	5
17	Tiliaceae	1	2
18	Linaceae	2	4
19	Geraniaceae	3	16
20	Ilicaceae	1	1
21	Celastraceae	1	1
22	Rhamnaceae	1	2
23	Aceraceae	1	1
24	Leguminosae	17	67
25	Rosaceae	14	63
26	Saxifragaceae	4	29
27	Crassulaceae	3	13
28	Droseraceae	1	3
29	Haloragaceae	3	11
30	Lythraceae	2	2
31	Onagraceae	3	16
32	Cucurbitaceae	1	1
33	Umbelliferae	34	56
34	Araliaceae	1	1
35	Cornaceae	1	2
36	Caprifoliaceae	5	6
37	Rubiaceae	4	18
38	Valerianaceae	2	7
39	Dipsacaceae	2	5
40	Compositae	42	114
41	Campanulaceae	6	12
42	Vacciniaceae	2	4
43	Ericaceae	9	19
44	Monotropaceae	1	1
45	Plumbaginaceae	2	11
46	Primulaceae	8	17
47	Oleaceae	2	2
48	Apocynaceae	1	1
49	Gentianaceae	7	20
50	Polemoniaceae	1	1
51	Boraginaceae	9	23
52	Convolvulaceae	3	5
53	Solanaceae	3	4
54	Scrophulariaceae	13	70
55	Orobanchaceae	2	12
56	Lentibulariaceae	2	10
57	Verbenaceae	1	1
58	Labiatae	18	58
59	Plantaginaceae	2	9
60	Illecebraceae	4	6
61	Chenopodiaceae	6	26
62	Polygonaceae	3	29
63	Aristolochiaceae	1	1
64	Thymelaeaceae	1	2
65	Elaeagnaceae	1	1
66	Loranthaceae	1	1
67	Santalaceae	1	1
68	Euphorbiaceae	3	14
69	Ulmaceae	1	5
70	Urticaceae	3	4
71	Myricaceae	1	1
72	Amentaceae	6	9
73	Salicaceae	2	19
74	Empetraceae	1	1
75	Ceratophyllaceae	1	2
76	Hydrocharitaceae	2	2
77	Orchidaceae	16	50
78	Iridaceae	4	5
79	Amaryllidaceae	3	4
80	Dioscoreaceae	1	1
81	Liliaceae	19	30
82	Juncaceae	2	30
83	Typhaceae	2	7
84	Araceae	1	2
85	Lemnaceae	2	5
86	Alismataceae	6	7
87	Naiadaceae	7	40
88	Eriocaulaceae	1	1
89	Cyperaceae	9	105
90	Gramineae	47	128
91	Pinaceae	3	4
	Totals :	468	1,513

with Gramineae third. The figures for Dicotyledones and Monocotyledones are :

Dicotyledones :	343 genera,	1,092 species
Monocotyledones :	122 genera,	417 species
Total :	465 genera,	1,509 species

This gives, on the basis of species, the Monocotyledones a 27·6 per cent representation in the angiospermous flora.

We turn now to a consideration of the distribution of plants within the British Isles. The geographical area occupied by a species (or other systematic group) is termed its range. This can be represented in various ways by marking records on outline maps. For the British Isles as a whole a much-used method involves the use of vice-counties. The counties of Great Britain, and separately those of Ireland, are accepted with certain modifications of their county boundaries, but the larger counties are further subdivided and some smaller ones added to others. The resultant scheme is easy to use for there is most often little or no difficulty in finding the vice-county for any given record. It is also easy to tabulate, summate, and map ranges on a vice-county basis. The limitations of the vice-county method must, however, be remembered. No account is taken of distribution within the vice-county—a common plant ranging over the entire area is equated with one occurring locally in a single small patch. The vice-county boundaries are mainly of historico-political origin and most often do not coincide with such natural features as influence the distribution of plants. The vice-counties are diverse in size and shape. The value of new vice-county records can be over-estimated.

Since the use of vice-counties is extending even beyond the realms of botany the scheme accepted for the NEW NATURALIST series is given in full on page 260. The original scheme was set out over 90 years ago by H. C. Watson.[1] It is now recognized as desirable that there should be clarification of certain discrepancies in usage and a final settlement of the exact boundaries of all the vice-counties. The matter is lucidly discussed in a recent paper[2] and a joint committee of the Systematics Association and the Botanical Society are investigating the problems involved and are preparing a full account of their findings for publication.

[1] *Cybele Britannica*, vol. 3, pp. 526-28 : 1852, and vol. 4, pp. 139-42 : 1859.
[2] *Bot. Soc. Exch. Club 1941-42 Report :* 524-26 : 1944.

There are two ways in which the distribution of British plants has been considered : (1) by using the vice-county system a classification has been made of ranges within Great Britain only (Ireland was not originally included) ; (2) by stressing the general distribution of British species in Europe these have been grouped into classes based more on continental than on British distribution. An entirely satisfactory classification combining both viewpoints has not yet been proposed and, indeed, cannot be expected before much more research has been done on the ecology, variations, and history of individual species and groups of presumably allied species. The schemes that have been used are of considerable interest and value and an outline of some of them must be attempted.

The following " range-types " are useful for classifying the seed-bearing plants within Great Britain :

1. BRITISH TYPE, i.e. species occurring more or less throughout Great Britain, such as :

alder (*Alnus glutinosa*)
hazel (*Corylus avellana*)
carrot (*Daucus carota*) (Pl. 5, p. 46)
hemlock (*Conium maculatum*)
primrose (*Primula vulgaris*)
bell heather (*Erica cinerea*)
cross-leaved heath (*Erica tetralix*)

2. ENGLISH TYPE, i.e. species occurring more or less widely in England but becoming rarer northwards and very rare or absent in Scotland. Examples are :

white bryony (*Bryonia dioica*)
black bryony (*Tamus communis*) (Pl. 3, p. 42)
autumnal squill (*Scilla autumnalis*)
foetid iris (*Iris foetidissima*)
dogwood (*Cornus sanguinea*)
yellow archangel (*Galeobdolon luteum*)
woolly thistle (*Cirsium eriophorum*)
green hellebore (*Helleborus viridis*) (Pl. 6, p. 47)
parsnip (*Pastinaca sativa*)
marsh gentian (*Gentiana pneumonanthe*) (Pl. 7, p. 62)
oxlip (*Primula elatior* auct.) (Pl. III, p. 10)
dark mullein (*Verbascum nigrum*) (Pl. I, p. 2)

3. SCOTTISH TYPE, i.e. confined to Scotland, or at least very rare in England, and including :

Scots pine (*Pinus silvestris* var. *scotica*)
globe-flower (*Trollius europaeus*)
wood crane's-bill (*Geranium silvaticum*)

4. INTERMEDIATE TYPE, i.e. plants which occur only in the north of England and in Scotland, with, as examples :

baneberry (*Actaea spicata*)
water lobelia (*Lobelia dortmanna*)
bird's-eye primrose (*Primula farinosa*)

5. LOCAL TYPE, i.e. species of very limited range confined to one or a few localities. Thus the Cheddar pink (*Dianthus caesius*) is limited to the Mendips, *Lloydia serotina* to Snowdon, and *Phyllodoce caerulea* to the Sow of Athol in Perthshire.

The English type includes two sub-types. Species occurring chiefly in the east of England have been called Germanic, with the oxlip (*Primula elatior*), pasque-flower (*Anemone pulsatilla*) (Pl. 4, p. 43), ground-pine (*Ajuga chamaepitys*), mousetail (*Myosurus minimus*), rampion (*Phyteuma tenerum*), snowdrop (*Galanthus nivalis*) (Pl. 2b, p. 35), grape-hyacinth (*Muscari racemosum*) (Pl. 2a, p. 35), and sea-heath (*Frankenia laevis*) as characteristic examples. Plants found mainly or entirely in western England and Wales (and sometimes in Ireland) are designated Atlantic, with English stonecrop (*Sedum anglicum*), tree-mallow (*Lavatera arborea*), pennywort (*Umbilicus pendulinus*), Welsh poppy (*Meconopsis cambrica*), a clover (*Trifolium bocconi*), and viscid bartsia (*Bartsia viscosa*) as examples. The term " Atlantic " has also been used by some, especially Continental authors, to designate species, or other taxonomic classes, confined to western or " Atlantic " Europe, though the conception of " Atlantic " Europe is not always the same.

Within the Scottish type a Highland sub-type can be more or less clearly distinguished, consisting of species limited to the mountain areas or their vicinity. This sub-type includes the trailing azalea (*Loiseleuria procumbens*), the alpine speedwell (*Veronica alpina*), the alpine cat's-tail grass (*Phleum alpinum*), and alpine forget-me-not (*Myosotis alpestris*) (Pl. 8, p. 63).

The altitudinal ranges of British seed plants show some interesting features. Our highest altitudes are in the west and north : Snowdon 3,560 ft. (1,085 m.), Ben Nevis 4,409 ft. (1,344 m.) etc. Such altitudes in themselves probably have very little direct effect on plant life, decrease of pressure and increase of insolation being too small. On the other hand their indirect influence through modifications of climatic factors (giving local climates) and on soil factors in the broad sense is considerable. On Ben Nevis there is an average lapse rate of temperature of 1° for every 270 feet. Rainfall is generally heavier on mountains than elsewhere and our western mountains get the full force of the rain-bearing winds from the south-west and west. The skies are often cloudy and the relative humidity frequently high. On the whole, in the west, the summers are cool and the winters mild. On the basis of altitude alone three classes can be conveniently made : Sea level (0–50 ft.), Lowland (up to 1,000 ft.), and Upland (above 1,000 ft.). Examples of species placed in these classes are :

SEA-LEVEL

Cochlearia anglica, a scurvy-grass.
Cakile maritima, sea-rocket.
Frankenia laevis, sea-heath.
Spergularia salina, a sand-spurrey.
Polycarpon tetraphyllum, four-leaved allseed.
Lathyrus japonicus, sea-pea.
Aster tripolium, sea-aster.
Mertensia maritima, sea-lungwort.
Beta maritima, sea-beet.
Juncus maritimus, sea-rush.
Spartina stricta, cordgrass.
Agropyron junceum, a couchgrass.

LOWLAND

Clematis vitalba, traveller's joy.
Anemone pulsatilla, pasque-flower.
Ranunculus fluitans, a water crowfoot.
Reseda luteola, dyer's rocket.
Viola odorata, sweet violet.
Silene conica, a catchfly.
Stellaria palustris, marsh stitchwort.
Ulex minor, dwarf gorse.
Prunus insititia, bullace.
Bryonia dioica, white bryony.
Cicuta virosa, cowbane.
Cirsium eriophorum, woolly thistle.

UPLAND

Thalictrum alpinum, alpine meadow-rue.
Draba rupestris, rock whitlow-grass.
Sagina linnaei, a pearlwort.
Astragalus alpinus, alpine milk-vetch.
Saxifraga nivalis, alpine saxifrage.
Gentiana nivalis, small gentian.
Veronica alpina, alpine speedwell.
Salix lanata, woolly willow (Pl. IV, p. 11).
Salix reticulata, reticulate willow.
Juncus trifidus, highland rush.
Juncus triglumis, three-flowered rush.
Carex rariflora, a sedge.

A large number of species overlap from one class to another and there are also some altitudinal ranges whose ecological basis has not yet been satisfactorily explained. The common heather (*Calluna vulgaris*) (Pl. 17, p. 90) is found from very near sea-level to 3,414 ft. A considerable number of plants typical of sea-level communities also occur in the mountains. Thus the sea bladder campion (*Silene maritima*) is recorded from 3,180 ft. in the Scottish Highlands, the sea thrift (*Armeria maritima* s.l.) from 4,160 ft. on Ben Nevis, and the sea plantain (*Plantago maritima*) from 2,600 ft. on Snowdon and in Breadalbane. In Ireland the London Pride (*Saxifraga umbrosa*) ranges, at least in a broad taxonomic sense of the species, from sea-level in Kerry and Mayo to 2,000 ft. in Donegal, 2,500 ft. in Tipperary, 2,600 ft. in Waterford, and 3,414 ft. in Kerry. The mountain avens (*Dryas octopetala* (Pl. 10b, p. 67) is most often an upland species, but descends to sea level in northern Scotland and in western Ireland (Galway), and its altitudinal extremes are sea-level to 2,700 ft. (in Perthshire). The red whortleberry or cowberry (*Vaccinium vitis-idaea*) is also mainly an upland species and is recorded from 3,430 ft. on Snowdon, but descends to below 200 ft. in several localities.

A number of figures for the Irish flora are of considerable comparative interest. The following table may require a number of additions but these would probably make little alteration in the percentages, unless a very different species standard than that used throughout this work were adopted :

Type	*Number of Species in England and Wales*	*Number in Ireland*
British	544	536 or 98·0% *of number in England and Wales*
English	381	246 or 64·5% ,, ,,
Scottish (& Intermediate)	98	62 or 63·3% ,, ,,
Highland	67	41 or 61·2% ,, ,,
Atlantic	62	34 or 55·0% ,, ,,
Germanic	102	12 or 11·7% ,, ,,

It is significant that all but 8 of the species widespread in Great Britain occur also in Ireland. On the other hand, 90, or 88·3 per cent, Germanic species present in Great Britain do not reach Ireland. Since these latter are restricted, more or less, to eastern England it seems very likely that ecological (especially climatic) factors and not the insularity of Ireland are mainly responsible for their absence from Ireland. This explanation may also apply to absentees of other types, but is least likely for many of the 28 Atlantic species present in Great Britain but not in Ireland. About 20 species of flowering plants are present in Ireland and not in Great Britain. The history of most of these is debatable and probably very varied, but the present restriction of range of many of them is almost certainly a question of tolerance, again especially of climatic conditions.

Considerable attention has recently been paid to the plant life of the islands off the western coast of Scotland. These numerous islands have very little woody vegetation, though it remains doubtful whether or not formerly existing pine forests largely covered them and were destroyed by man. Most of the species so far recorded have wide ranges in the British Isles but particular interest attaches to the occurrence of some which are usually regarded as "arctic-alpine" (see p. 61), such as northern rock-cress (*Arabis petraea*), moss campion (*Silene acaulis*), mountain avens (*Dryas octopetala*) (Pl. 10b, p. 67), alpine hair-grass (*Deschampsia alpina*), mountain sorrel (*Oxyria digyna*) (Pl. 9, p. 66), and a bistort (*Polygonum viviparum*) ; some of very local distribution elsewhere in the British Isles, as *Juncus pygmaeus* (Lizard), *J. capitatus* (Land's End, Lizard, Anglesey, and Channel Islands), *Cicendia pusilla* (Channel Islands), *Thlaspi calaminare* (Derbyshire), and *Arenaria norvegica* (Sutherland and Shetland) ; and a number of microspecies, or units of a major species, limited to the Scottish islands or with the islands apparently the centre of their ranges, as *Orchis hebridensis*, *Euphrasia campbellae*, and some fescue-grasses and hawk-weeds. The peculiar and discontinuous distribution of some elements of the flora led to the suggestion that many of the species concerned had survived the Ice Age in or near their present locations.

The natural phanerogamic flora of the isolated island of St. Kilda consists of aquatic and marsh species, plants of sea-cliffs and rocks, a small number of arctic-alpine species, a few possible relicts of or migrants from a woodland ground flora, and, in main bulk, heath-moor species. The heath-moor vegetation is dominant and can be accepted as the natural climax community for the greater part of the island. It is to be regarded as relatively old and it may well have survived the Ice Age more or less *in situ*. Certainly the floristic composition and the general ecological grouping suggest that St. Kilda's plant cover is a detached piece of West Scottish vegetation.

One example of a northern island may be given. Foula is the most westerly of the Shetland (or Zetland) group and attains an altitude of 1,373 ft. (418 m.). The vegetation is dominantly of heath-moor type and the natural flora has been classified into the following groups on local habitat conditions : aquatic and marsh, sea-cliffs, rocks and sea-strand, arctic-alpine (very few), relicts of woodland ground-flora or migrants from woodlands (very few), heath-moor (the main bulk). In the Ice Age the main ice-sheet reached the island but did not completely over-ride it. As with St. Kilda, the greater part of the flora and vegetation of Foula very possibly survived in the island in places free from ice, though the arctic-alpine element may have been more important then than it is now. Tree species, especially birch, came in from the south as conditions became more favourable, the arctic-alpine element became reduced, and later the advent of man, the introduction of sheep, and, perhaps, increase of wind storms, destroyed all, or nearly all, traces of natural woodland.

CHAPTER 7

THE GEOGRAPHICAL RELATIONSHIPS OF OUR FLORA

WHEN WE ATTEMPT a classification of British plants largely on the basis of the distribution of the species outside Britain, and mainly in the near-lying parts of Europe, we find a difficulty in deciding how many categories or classes to make. Several schemes have been proposed recently and the following outline attempts to combine the best of the more important in a simplified manner by making nine main classes. While the groups as a whole represent types of distribution which must be accounted for in any discussion of the origin of the British flora, a certain amount of overlapping occurs and sometimes intermediate groups could be formed. Moreover, some of the groups are rather doubtfully distinct and may well not represent elements of distinct migrations into the British Isles. Thus our groups (4) Western European (oceanic) and (5) Southern (including " Mediterranean ") could reasonably be considered as forming a general " southern element " so far as the British flora is concerned. Occasionally a species, though essentially a member of one group, extends its range under special conditions into the general area of another group. It also not infrequently happens that a species which is quite correctly classified as a member of a certain group by no means occupies the whole geographical area by which the group is designated. Thus a " Mediterranean " species does not necessarily occur in every part of the Mediterranean Region.

1. *Widely distributed species.* These are species with a wide distribution in the Northern Hemisphere or at least in the Eurasian continent. Some are almost cosmopolitan, occurring in the Northern and Southern, Eastern and Western Hemispheres. Obviously the group could be much subdivided but this is not necessary for our present purpose. The group is a large one in the British flora and consists of nearly 700 species. Selected examples amongst well-known plants are : the common floating pondweed (*Potamogeton natans*), recorded from Europe, N. Asia, India, Africa, N. America, and Australia ; a sedge, *Carex pseudocyperus*, known from Europe, N. and S. Africa, Asia,

America, and Australia ; sweet gale (*Myrica gale*) in W. and N.W. Europe, N. Asia, and N. America ; white deadnettle (*Lamium album*) in Europe, N. Africa, and N. Asia, and introduced into N. America and elsewhere ; water figwort (*Scrophularia aquatica*) occurring in Central Europe, N. Africa, N. and W. Asia, and the Himalayas ; grass of Parnassus (*Parnassia palustris*) (Pl. 10a, p. 67) distributed in Europe, N. Africa, N. and W. and Central Asia, and N. America ; mousetail (*Myosurus minimus*) known from Europe, W. Asia, N. and S. Africa ; yellow water-lily (*Nuphar luteum*) in Europe, temperate Asia, and N. America ; common poppy (*Papaver rhoeas*) occurring in Europe, N. Africa, and W. Asia to India ; bulbous buttercup (*Ranunculus bulbosus*) in Europe, Asia, and N. Africa, introduced in America ; and Scottish thistle (*Onopordum acanthium*) (Pl. 11, p. 74) in Europe and temperate Asia. A large number of water plants (such as the pondweed mentioned above) have a particularly wide distribution. There are also many of our common weeds of arable land, like the common poppy, which now have a very wide range. It is, indeed, often difficult or impossible to determine the original natural range of many weeds and ruderals or even to say how, whence, and when they reached our country. While, most often, plants of our flora with a wide general range also have a wide range in the British Isles, there are exceptions. For example, the water plant *Naias marina*, with a wide distribution in the temperate and tropical regions of the Old World, is known in Britain only from three of the Norfolk Broads.

2. *General European species.*

(*a*) Species occurring in all or most parts of Europe. This is only a moderate-sized group of species, about 130 in number, which are absent from or only just enter Asia, though a small number of them (some 25) reach N.W. Africa. Some of the members of this category are absent from the extreme north and east of Europe.

Examples are :

globe flower (*Trollius europaeus*)
traveller's joy (*Clematis vitalba*)
wood anemone (*Anemone nemorosa*) (Pl. 26, p. 111)
meadow rue (*Thalictrum minus*)
musk mallow (*Malva moschata*)
slender clover (*Trifolium filiforme*)
hairy greenweed (*Genista pilosa*)
strawberry-leaved potentil (*Potentilla sterilis*)

cat's-ear (*Hypochoeris radicata*)
creeping jenny (*Lysimachia nummularia*)
yellow pimpernel (*L. nemorum*)

(*b*) High mountain species in Central (and sometimes southern) Europe, occurring also in the Arctic Region. This group, of about 76 species, consists of plants most often more or less circumpolar in their distribution with present ranges, often discontinuous, including high mountains to the south of the Arctic, especially the Alps and Pyrenees. This group has been termed the Arctic-Alpine element. In Britain the altitudinal limits of the species are very variable, but most of them do not descend below 600 m. Interesting exceptions to this statement are some " Arctic-Alpine " species which occur near sea-level in the north of Scotland and in western Ireland. As might be expected, most of the species of this group occur in the north and west of both Great Britain and Ireland, and above all in the Scottish Highlands—63 are recorded from mid-Perth. In England, about half are found in Westmorland. Examples of our " Arctic-Alpine " element are :

alpine lady's mantle (*Alchemilla alpina*)
dwarf willow (*Salix herbacea*)
dwarf birch (*Betula nana*)
moss campion (*Silene acaulis*)
alpine meadow-rue (*Thalictrum alpinum*)
highland rush (*Juncus trifidus*)
purple saxifrage (*Saxifraga oppositifolia*)
small gentian (*Gentiana nivalis*)
alpine willow-herb (*Epilobium alpinum*)
bog whortleberry (*Vaccinium uliginosum*)
mountain azalea (*Loiseleuria procumbens*)

3. *Central European species* (*continental*). The flora of Central Europe is itself heterogeneous in origin, and its complicated history has by no means been completely unravelled. Here we are concerned with three sections which are represented in our British flora :
(*a*) General Central European species are about 70 in number. They are exemplified by :

hornbeam (*Carpinus betulus*)
wych elm (*Ulmus glabra*)
milkvetch (*Astragalus glycyphyllos*)

green hound's-tongue (*Cynoglossum germanicum*)
fritillary (*Fritillaria meleagris*) (Pls. 34, p. 131, and 35, p. 138)
dwarf orchid (*Orchis ustulata*)
great spearwort (*Ranunculus lingua*)
grape hyacinth (*Muscari racemosum*) (Pl. 2a, p. 35)
yellow archangel (*Galeobdolon luteum*)

(*b*) Steppe plants constitute a difficulty because of the diverse views which have been expressed as to what should be the limits assigned to the steppe flora. If a broader view be taken quite a number of the species included with the 70 general Central European species may be included, but taking a narrow view there are only about 10 British plants which can best be classed as steppe plants, and these are mainly found in East Anglia, particularly in Brecklands. Examples are: a catchfly (*Silene otites*), spiked speedwell (*Veronica spicata*), field ragwort (*Senecio campestris*), field wormwood (*Artemisia campestris*), and a cat's-tail grass (*Phleum boehmeri*).

(*c*) High mountain species form a small group of about 9 species which occur in the Central European mountains and reach their northern limits in the British Isles, not extending into the Arctic. They are sometimes referred to, not very exactly, as the "Alpine" element in our flora and are exemplified by: British rock-cress (*Arabis stricta*), cyphal (*Arenaria sedoides*), spring gentian (*Gentiana verna*), cut-leaved saxifrage (*Saxifraga hypnoides*), and alpine penny-cress (*Thlaspi alpestre*). These plants on the Continent are found usually at higher altitudes than in the British Isles (several of them ascending to 3,400 m. or over) and are thus in accord with a general rule that high mountain species are found with ranges at lower altitudes as one goes from south to north in the Northern Hemisphere.

4. *West European species* (*oceanic*) contrast with the Central European in that their ranges are within the area of Atlantic climate. Some 75 species may be referred to this element though some have a wider and some a more restricted general distribution. Very typical of this group are our three species of gorse (*Ulex europaeus*, *U. gallii*, and *U. minor*). A number of heaths must also be specially mentioned (*Erica ciliaris*, *E. tetralix*, and *E. vagans*).

Other characteristic examples are:

hemlock water dropwort (*Oenanthe crocata*)
ivy-leaved bellflower (*Wahlenbergia hederacea*)

PLATE 7

Brian Perkins

MARSH GENTIAN, *Gentiana pneumonanthe* (Gentianaceae)
Chobham Common, Surrey. August 1946

PLATE 8

Brian Perkins

MOUNTAIN FORGET-ME-NOT, *Myosotis alpestris* (Boraginaceae), Ben Lawers, Perthshire, June 1946

sea kale (*Crambe maritima*)
sea radish (*Raphanus maritimus*)
bluebell (*Scilla non-scripta*) (Pl. 24, p. 107)
ivy-leaved crowfoot (*Ranunculus hederaceus*)
bog St. John's wort (*Hypericum elodes*)

5. *Southern species* (including the so-called Lusitanian plants), as here accepted, can be classified into two sub-sections:
(*a*) Species of southern Europe, excluding the Mediterranean Basin proper, but often with ranges overlapping those of Mediterranean species and/or of West European species. Examples are:

sea beet (*Beta maritima*)
navelwort (*Umbilicus pendulinus*)
rock samphire (*Crithmum maritimum*) (Pl. 13, p. 78)
ivy broomrape (*Orobanche hederae*)
sea lavender (*Limonium vulgare*)
tutsan (*Hypericum androsaemum*)
small-flowered crowfoot (*Ranunculus parviflorus*)
sea spurge (*Euphorbia paralias*)

(*b*) Mediterranean species in the restricted sense of having their main range in countries of the Mediterranean Basin. These include:

strawberry tree (*Arbutus unedo*)
golden samphire (*Inula crithmoides*)
matted sea lavender (*Limonium reticulatum*)
seablite (*Suaeda fruticosa*)
bladder seed (*Danaa cornubiensis*)
nitgrass (*Gastridium lendigerum*)
tree mallow (*Lavatera arborea*)
Romulea columnae
wild madder (*Rubia peregrina*)

6. *Northern species*. As a general group of about 30 members, they are conveniently divided between two subsections:
(*a*) Arctic and subarctic species, including a number of sedges (*Carex grahami, C. rariflora, C. sadleri, C. recta, C. saxatilis*), sea lungwort (*Mertensia maritima*), cloudberry (*Rubus chamaemorus*), alpine pearlwort (*Sagina nivalis*), Scottish bird's-eye primrose (*Primula scotica*), and dwarf cornel (*Cornus suecica*).
(*b*) Northern, especially Scandinavian, species number well over 100 in our flora, if, as is done here, some with outlying discontinuities of

range in Central Europe be also included. Examples are : a considerable number of sedges (such as *Carex canescens*, *C. disticha*, *C. limosa*, *C. montana*, and *C. pulicaris*), four winter-greens (*Pyrola media*, *P. rotundifolia*, *P. secunda*, and *Moneses uniflora*) (Pl. XIII, p. 194), bogbean (*Menyanthes trifoliata*), mountain cat's-foot (*Antennaria dioica*) (Pl. 12, p. 75), the creeping goodyera (*Goodyera repens*), the linnaea (*Linnaea borealis*), and holy grass (*Hierochloe borealis*).

7. *American species.* There are, at most, about half a dozen species which have a more or less wide distribution in N. America but in Europe occur only in the British Isles. The best-known examples are pipewort (*Eriocaulon septangulare*), blue-eyed grass (*Sisyrinchium angustifolium*, a member of the iris family and botanically not a grass), and Irish lady's tresses (*Spiranthes romanzoffiana*).

8. *Species whose centre of distribution is the British Isles.* The best example is undoubtedly the rice grass (*Spartina townsendii*) (Pl. 48, p. 223) whose history is outlined later on in this work (see p. 243).

9. *Endemic British species*, those known to occur only in the British Isles, are mostly not very distinct from more widely distributed species (i.e. they are " small " species or microspecies, which are regarded as subspecies or varieties by some botanists). It is doubtful if more than 15 to 20 endemic species can be accepted in the British flora (seed-bearing plants). Among the best characterized are : a monkshood (*Aconitum anglicum*) from south-west and west England and east Wales; a fumitory (*Fumaria occidentalis*), found only in Cornwall ; a cabbage-like plant (*Brassicella wrightii*) known only from Lundy Island ; two sea lavenders (*Limonium recurvum* and *L. transwallianum*) from Dorset and Pembrokeshire respectively; a forget-me-not (*Myosotis brevifolia*) ; and a brome grass (*Bromus interruptus*).

It must be emphasized that in the above scheme certain " critical " genera (or sections of genera) have been entirely ignored. This is not because they are regarded as unimportant but because they open up special problems which are dealt with in later chapters and because our knowledge regarding the systematic value and distribution of many of them is far from complete. Such genera include : *Rosa*, *Rubus*, *Hieracium*, *Taraxacum*, *Sorbus* (sometimes regarded as a section of *Pyrus*), *Alectorolophus*, and *Euphrasia*.

South-east England is the part of the country nearest to the Continent. It is, therefore, interesting to consider the results of a comparison of the flowering plants of Kent with those of the Pas de Calais. It has been shown that 54 species occur in Kent but not in the Pas de Calais, while 88 species are recorded for the Pas de Calais but not for Kent. The conclusions drawn from the study are that the English Channel, at least at the eastern end, has been in itself unimportant as a barrier to plant migration and that all but perhaps a very minute proportion of the striking differences between the floras of Kent and Pas de Calais are due to differences in external conditions in the two regions and not to lack of dispersal. The general suggestion is made that the external factors controlling plant-range are far more delicately balanced than is generally supposed. Unfortunately, there has been no detailed or experimental consideration of the ecology, and still less of the past history, of the 142 species peculiar to either Kent or the Pas de Calais and it may well be that many of the localized species are fairly new arrivals, particularly from the west, to whose migration the Straits of Dover constitute perhaps a temporary barrier. This suggestion is not intended to minimize the importance of ecological factors.

CHAPTER 8

THE HABITAT FACTORS OF BRITISH PLANTS

IN DEALING with the plant life of any area and in trying to understand its composition and its history it is essential to consider not only the plants themselves but also the conditions under which they live. The surroundings, frequently termed the external environment, of either an individual plant or of a community of plants are always complex. In order to investigate the action of the external environment on plant life we have often to analyse it into component parts and to observe the effects of, and to experiment with, a factor which varies or is varied in intensity while, as far as possible, other factors are kept constant. This necessary procedure has its dangers. No factor ever acts in isolation from other factors and the " habitat " is the place of abode of a plant together with a summation of all factors acting on the plant from outside the plant itself. From time to time it is highly desirable to recall the effective unity in action of the habitat factors and to remember that the plant's reaction is to their integration. Nevertheless, one factor may often appear to have a predominating influence because, in some way or another, it deviates markedly from the average or from some optimum. In experimental research with plants a further difficulty arises. The more natural are the conditions under which a plant is grown, the less are the habitat factors under control and the less accurately can they and the plant's reactions to them be measured. Conversely, the greater the degree of experimental control, the more artificial becomes the external environment to which the plant is subjected and often one may well wonder how far the results obtained throw light on the problems of plant life under natural conditions. It is, therefore, essential constantly to check experimentally obtained results with behaviour as seen in nature before generalizations are made.

Habitat factors can be classified in various ways. For our present purpose the following classification will be most useful :

A. *Inorganic*
   1. Climatic
   2. Edaphic (soil)

B. *Organic*
   3. Biotic (animals and plants)

PLATE 9

Brian Perkins

MOUNTAIN SORREL, *Oxyria digyna* (Polygonaceae). Ben Lawers, Perthshire. June 1946

B. A. Crouch

a. GRASS OF PARNASSUS, *Parnassia palustris* (Parnassiaceae)
Lower slopes of Yewbarrow, Wasdale, Cumberland. September 1945

J. S. Barlee

b. MOUNTAIN AVENS, *Dryas octopetala* (Rosaceae). Black Head, Co. Clare. September 1945

These groups of factors (for none is an indivisible unit) interact in numerous ways. It is also necessary to explain why certain topographic or physiographic variations are not considered as basically distinct factors or factor-groups. Increasing altitude, for example, is frequently correlated with changes in flora and vegetation, but in its action on plant life altitudinal variation acts through changes in the climatic, edaphic and, usually to a less conspicuous degree, biotic factors. Thus temperature decreases, precipitation often increases, insolation increases and changes in quality with increasing altitude, while soils may become shallower, differently drained, and differently composed the greater the height above sea-level, and the species of insect pollinators and animals dispersing fruits and seeds change. It is frequently desirable, in dealing with a mountainous area, primarily to record the effects of altitude on the flora and vegetation as if the cause of the variations—usually strongly zonal in character—were a single factor scorable in metres or feet. When data are available, however, altitude is always found to be analysable into the three primary factor-groups and their interactions. Much the same holds for distance from the sea, degree of slope, or any other physiographic or topographic causes of floristic or vegetational changes.

We have already seen the interest added to the study of existing plant life if its history be taken into account. Indeed, to understand the present composition and relationships of a flora its history must be investigated. In dealing with the British flora, for example, its major natural changes, so far as they are known, are obviously correlated with great climatic changes which occurred in the course of geological time. Small floristic changes such as are always occurring are correlated with, and often in some sense due to, changes in the kind or intensity of small climatic, edaphic, or biotic factors. We shall have still further evidence in the next chapter that vegetation also constantly changes over short, medium, or long periods. We have, therefore, to envisage a historical aspect to all three factor groups and to their actions and interactions. With this historical aspect is closely associated what may be termed the geographical aspect. In the flora and vegetation of a country there will be only such species of plants as have either evolved there (probably very few indeed in Britain) or have reached the area from outside and have become established. We know that many species which can flourish in habitats occurring here are not constituents of our natural flora and vegetation because they could not get here by natural means, for historico-geographical reasons. Floristic or vegetational composition depends to an unknown but surely considerable degree on availability of supply of immigrants.

In considering the environmental factors in this chapter two aspects will be kept to the fore: the selective influence of the factors on the flora and their importance in limiting the development, composition, and physiognomy of plant communities. Illustrations will therefore be taken from both floristics and ecology.

Climatic Factors

The climatic factor-group is in one sense the most important of the three groups. The world zones of temperatures are well known, and closely correlated with them are different floras and plant communities. Moreover, climate acts, more or less directly, on the soil and biotic factors. On the other hand, every habitat has what is termed a micro-climate which may affect a very small area indeed and is often in part determined by vegetation. Thus, in a wood different layers below the dominant overhead tree canopy have significant even if small differences in temperature, moisture, and light supply.

Climatic factors may be considered under the following heads: temperature, precipitation, humidity, light, and wind. To understand the importance of these for the plant life of the British Isles it is necessary to outline both their records for the area and their actions on plants. Records of climatic factors can be kept and utilized in various ways. Their influence on plant life can only be shown fully by reference to both averages and extremes and always by remembering the complications of their interactions. It may also be noted that the position and distribution of meteorological stations are not always such as the botanist would desire.

Temperature. Temperature is recorded by means of thermometers or thermographs (self-registering thermometers), of which there are many varieties. In utilizing data provided by meteorologists it is important to remember that both temperature and pressure are usually reduced to sea-level before they are plotted on charts, i.e. for temperature 1° F. is added for every 300 feet in the elevation of the station or 1a (= 1° absolute) for every 165 m., while other rates are used for maximum and minimum temperatures. This procedure is desirable for certain purposes but at high altitudes may give a false idea of the actual temperature conditions especially as biological factors.

The British Isles lie within the North Temperate Climatic Zone. Their position on the Atlantic edge of Europe and the influence of the prevalent "westerlies" together with the warmer ocean drifts (usually classified as a continuation of the Gulf Stream) give them a more equable, oceanic type of climate than many other lands in approximately the same latitudes. There are not the extremes of summer heat or of winter cold of many North Temperate continental land masses or even of some coastal districts. The southern part of Ireland and the south-west of England and Wales lie in the northern part of the 60°–50° F. average annual isotherm belt (belt of equal temperatures) and the remainder within the 50°–40° F. belt. Within the British Isles the two factors most influencing temperature are distance from the sea, coastal districts being warmer, and altitude, with increase of which temperature decreases. The average "lapse rate," that is the decrease of temperature with increase of altitude, may be taken as 1° F. for 280 ft. vertical rise (6·5° C. per km.). It is topography more than latitude that makes the northern and north-western parts of Britain on the whole colder than the southern parts. There are exceptions to any generalization and it is well known that sheltered and otherwise favoured areas on the west coast of Scotland and some in west Ireland are amongst the most favourable to plant growth in the British Isles. On the other hand, deviations from the mean in any particular year for a given station are most frequently slight. The, usually fairly regular, seasonal variations from warm summer to cool or rather cold winter are of greater importance than mere annual averages.

Temperature as it affects the plant cover depends more or less directly on the sun but indirectly also on both edaphic (soil) and biotic factors. Within our latitudinal range there is considerable difference between the temperatures of summer and those of winter but no place in the British Isles is too hot or too cold for some plant growth. Every plant function has temperature limits above or below which it ceases, sometimes very abruptly, but these limits vary from species to species and even from variety to variety. This gives a wide field within which selection acts on physiological characters which are often, but not necessarily always, correlated with obvious structural features.

Respiration (the release of energy through oxidation) and photosynthesis (the manufacture of food by the green colouring matter) are directly affected by temperature changes but they are less easy to measure under natural habitat conditions than growth, and to a considerable degree growth summates the physiological activities of a plant. Growth will only occur above a minimum temperature. Apical growth has been found to be most rapid with a soil temperature at 31·3° C. for mustard (*Brassica alba*) and 25·6° C. for timothy grass (*Phleum pratense*). In many arctic and high mountain plants a very few degrees' rise above average winter temperature will stimulate them to growth. Some will grow under a foot of winter snow and produce new leaves or their seeds will germinate under snow. This is not so remarkable as it may at first seem because the snow cover acts as a protective blanket against both low temperatures and cold drying winds. The effects of hoar-frost on vegetation require investigation. Transpiration is affected by temperature both through the rate of water evaporation and the opening and closing of stomata, but depends even more on the relationship between temperature and atmospheric humidity which is considered below.

In the British Isles, temperatures are never sufficiently high directly to cause the death of plants, with a few possible exceptions as on almost bare rocks, and it is cold which is the most important limiting factor under the heading of temperature. The chemico-physical nature of death from cold is not fully understood. The vague statement that the nature and behaviour of the protoplasm determines the limits of low temperature from which given cells can recover is not much help towards a solution. It has been suggested that mechanical damage to the cells by ice crystals, that withdrawal of water from cells leading to their desiccation and alteration of the protoplasmic proteins, and that thawing causing flooding of the sap vacuoles, are the links between freezing and death of or injury to plants. It is of interest to note that a majority of the plants retaining green foliage during the winter (winter-green plants) have, at least during the cold months, sugar and not starch as their main carbohydrate food reserve. In a prolonged frost the low temperature of the soil, reducing the intake of water through the roots, can cause injury or death to the plant from drought rather than from cold as such. There is no doubt that severe late spring frosts do far more damage than even hard winter frosts, as witness the frost of 16–17 May, 1935. Cold air, if there be no wind, moves downwards, since it is heavier than warm air, and tends to collect at lower levels. This phenomenon of "frost-pockets" or "frost-hollows," in which temperatures are often lower than at higher and more exposed places, is well known to gardeners but its local influence on native vegetation has been very little studied. An interesting paper on the meteorological aspects of "frost-hollows" will be found in *Weather I*, No. 2, 41–45. A certain degree of cold can act as a stimulus to plant growth and development, and may even be obligatory, as in the germination of certain seeds. Gardeners are well aware that plants can, within limits, become "hardened" and know the best way of hardening individuals of many species. There is no doubt that, whatever the details of reaction to low temperatures, and especially to frost, may be, reduced temperatures are very important in limiting the spread of certain species northwards.

Precipitation. This occurs principally as rain or snow and is measured by rain-gauges, the automatic self-registering rain-gauge being termed a hyetograph. Water is essential to plant life but the more drastic limiting effects of low water supply cannot be studied on a large scale in the British Isles because our average precipitation is adequate in quantity and seasonal distribution for most plants within the limits of temperature and evaporation which normally occur. We have no deserts and even the effects of unduly prolonged summer droughts, such as that of 1921, are soon obliterated because they are exceptional. In the British Isles rainfall is greatest to the west and north and lowest in the east and south. This is owing to the highest ground being in the west and north with consequent interception of rain-bringing clouds which come mainly from the south-west. Differences in quantity and in deviations from annual means are much greater with rainfall than with temperature. The following table gives the annual rainfall in inches as the average of 30 years for a number of stations in Great Britain:

Shoeburyness	19·8	Holyhead	34·9
Cambridge	21·8	St. Ann's Head (Pembs.)	35·2
Kew	23·8	Plymouth	36·2
Spurn Point	24·2	Glasgow	37·2
York	24·3	Douglas	41·2
Tynemouth	24·5	Cardiff	41·4
Gorleston (Yarmouth)	24·5	Falmouth	43·6
Oxford	24·5	Stoneyhurst	46·6
Portland Bill	25·1	Buxton	48·4
Birmingham	26·5	Rothesay	49·0
Portsmouth	27·7	Stornoway	49·9
Liverpool	27·9	Fort William	77·8
Aberdeen	29·5	Seathwaite	129·5
Scilly Isles	31·9		

Extreme rainfall, recorded in one year, in inches, is :

	England and Wales	*Scotland*	*Ireland*	*British Isles*
Driest	24·66 (1921)	40·26 (1887)	33·34 (1887)	31·89 (1887)
% of average	70	80	77	77
Wettest	50·73 (1872)	67·43 (1872)	55·42 (1872)	56·73 (1872)
% of average	144	134	128	137

Total annual rainfall does not necessarily give information of much value to the plant ecologist. The seasonal distribution is often more important and for many purposes is best considered as monthly averages and extremes. In our islands, rainfall is fairly well distributed throughout the year. Temperature conditions are most favourable, in our latitudes, to plant growth in the summer with its longer days, and since rain is, on the average, also sufficient, growth is much greater then than in the winter. The seasonal interaction of factors is of very great importance and is well exemplified by the relationship between temperature and precipitation. Winter is here a rest period for plants ; summer is a rest period where a Mediterranean climate with summer drought holds sway. We have relatively few evergreens in our flora and wintergreen species are usually inhabitants of specially sheltered habitats like woods.

The greater rainfall to the west and north is, in part, responsible for the general western distribution of some Atlantic species in the British Isles, and conversely, the lower rainfall of the eastern counties, in part, accounts for the occurrence of continental or steppe species. One also notes the wider distribution of bogs and bog vegetation in the north and west. Probably the most striking differences due to rainfall, combined with atmospheric humidity, are in the relative richness of the bryophytic flora and vegetation. Mosses and liverworts are far more abundant in species and individuals and show much more lush growth in the north and west than in the south and east.

Snow affects plant life mainly by acting as a winter cover, thus affording some protection from prolonged low temperatures and from drying winds, and by causing some mechanical damage especially to evergreen trees and shrubs. The latter is much more obvious in parks and gardens where exotics are cultivated than amongst our native or well naturalized woody plants. Many readers will recall the exceptional "ice storm" of 27 to 29 January, 1940, when sleet froze on twigs and branches and caused considerable injury to a wide range of woody plants.

Atmospheric humidity. This is important because of its influence on evaporation of water from the soil, its deposition as dew, and its influence on the rate at which plants lose water through transpiration. Since at higher temperatures more water vapour can be taken into the air than at lower temperatures, the absolute quantity present at a given time is less important than the relative humidity, which is the amount of water vapour present in the air expressed as a percentage of the total amount of water vapour required to saturate it at the given temperature. Other conditions being equal, a low relative humidity means a high evaporation and a high transpiration rate. Humidity is measured by hygrometers (or psychrometers) of various patterns and may be continuously recorded by a self-registering instrument termed a hygrograph. The use of a hygrograph shows clearly the very considerable fluctuations in relative humidity which occur daily and seasonally at stations in the British Isles. The usually very regular rise at nights appears not to have been taken into account sufficiently in the study of 24-hour physiological rhythms in plants. There is, indeed, an interesting field for physiological-ecological research in the study of atmospheric humidity in relation to the distribution of plants and plant communities. A factor which exerts a direct action on such an important function as transpiration (the loss of water vapour) must play a limiting part in the composition of the flora and vegetation in any habitat. The highly humid air within a lowland oak forest or in a sheltered ravine seems to be directly correlated with the occurrence of delicate and apparently moisture-loving species and communities. Locally, vegetation itself in part controls relative humidity by giving off water vapour and by protecting from drying winds and from direct heat of the sun. It has, indeed, been shown that for evaporation and relative humidity the differences are much greater between sunny and shady than between otherwise dry and damp habitats. A dry atmosphere is known to affect the structure of plants in the course of their growth. For example it has experimentally been shown to restrict growth in height, increase rigidity of stems, diminish the length but increase the number of internodes, reduce size of leaf, increase thickness and deepen colour of leaf, promote development of hairs and stomata (breathing pores), increase development of root, hasten flowering and fruiting, augment glands, nectaries, and secretions in general, diminish diameter of epidermal (outermost skin) cells, reduce cortex (outer part of stems) and pith, favour development of bark and wood, and cause increase in number and hardness of spines and prickles.

Light. Sunlight is one of the essential factors in the manufacture of basic carbohydrate food in plants. Ultimately all living organisms, excepting only a few groups of Bacteria, depend upon photosynthesis or carbon assimilation, as the process of food-making by the green colouring matter or chlorophyll utilizing the energy of sunlight is termed.

Duration of sunshine is recorded at meteorological stations by special instruments which either concentrate the sun's rays in such a manner as to cause charring on a card or act on the principle of a "pinhole" camera and leave a trace on sensitized paper. Diffused light, however, is of great ecological importance and the measurement of light intensity has been much discussed. Photo-electric devices are undoubtedly the most accurate but the instruments are expensive and often rather cumbersome. A great deal of valuable information can be obtained by the careful use of a Watkins Bee Meter such as is used in photography. Usually the time taken for a small patch of photographic print to darken to a standard when exposed in a given habitat position is compared with the time required in full sunlight. Numerous exposures can be quickly made and with very simple precautions the results are comparable and instructive within certain limits. The main theoretical objection is that the rays affecting the photographic paper may be qualitatively different from those of most importance in controlling plant reactions. The meter cannot be used under water.

In addition to the photosynthetic action of light it has other effects on plants. Generally growth in length of stems is retarded by light. Plants grown in the dark or in deep shade have long slender stems and reduced leaves and are of a pale yellowish colour. They are said to be etiolated. The stomata of some plants open in light and close in the dark. Light affects the distribution and activity of growth hormones, or substances which influence rate of growth. Generally, the green colouring matter (chlorophyll) of aerial parts is only produced in light and, within limits, the amount produced, roughly estimated by the

F. Ballard

SCOTTISH THISTLE, *Onopordum acanthium* (Compositae). Near Sturdy's Castle, Woodstock, Oxfordshire. July 1945

John Markham

MOUNTAIN CAT'S-FOOT, *Antennaria dioica* (Compositae)
Cairngorms, Inverness-shire–Banffshire. June 1946

depth of colour, increases with the amount of sunlight. Roots and the hyphae (or threads) of some fungi also grow more quickly in the dark. Some seeds germinate only after exposure to light (e.g. species of rushes). Flower pigments are generally brighter and deeper as the insolation is more intense.

An interesting study of the effects of light on leaves can be made on any good-sized tree standing isolated. On the north side and on shaded branches the leaves will be " shade leaves " ; on the south side and on branches well exposed to a maximum amount of light there will be " sun leaves." The reader should examine a series of such trees of different kinds and record the differences in size, shape, margin, thickness, texture, venation, and (with a good lens or a microscope) internal structure of leaves in the different positions.

Differences of light intensity are of particular importance ecologically: (*a*) with reference to local topography, as on different sides of a hill or in different aspects of a valley ; (*b*) with different layers (strata) of a plant community, as the tree, shrub, herb, and moss layers in a wood ; and (*c*) with different depths of water, whether in the sea or in freshwater lakes. In general, within the British Isles variations in light intensity are only of local importance, i.e. they come within the realm of micro-climates. An exception to this generalization is the phenomenon known as photoperiodism. The relative length of day and night during the growing season—late spring and summer in the British Isles for most plants—depends on the latitude. It is found that plants react more particularly in their flowering and fruiting to length of day and on this basis can be classified as long-day plants, short-day plants, and neutrals. Typical North Temperate plants come into the first class, tropical plants into the second, and many widely distributed weeds into the third. It is probable that the differences in length of a summer day between the extreme south and the extreme north of Great Britain are sufficient to affect flowering of some species even of our native plants. Certainly we have here a factor, acting physiologically, which plays a part in controlling distribution of species and composition of floras in a world view.

Length of summer day

(Times from sunrise to sunset, for 21 and 22 June, 1946)

Scilly Isles, 50° N. — 16 hours 23 minutes.
Shetlands, 60° N. — 18 hours 53 minutes.

In high mountain regions the difference in both quantity and quality of the sunlight which reaches the lowlands and the peaks respectively has an influence on plant structure and survival. It is, however, doubtful if the mountains are high enough for this to have measurable ecological effects in this country. On the other hand the smoke palls which pollute the atmosphere in industrial areas do affect the vegetation. Sunlight is much reduced locally, leaves are shrouded, breathing pores are blocked with sooty deposits, and the rain is acidified. Some effects are noticeable at considerable distances from the centre of pollution. Lichens are particularly sensitive to such atmospheric pollution and a trustworthy indication of a pure atmosphere is an abundant lichen flora.

Wind. The intensity of wind is measured by anemometers and by expensive self-registering anemographs. The direction is also carefully recorded at meteorological stations. The effects of wind on plants are both direct and indirect. Heavy gales break branches and uproot trees and shrubs and may cause some mechanical damage to tall herbs. The indirect effects are more important and mainly act through the humidity factor by drying out the soil and increasing the loss of water vapour from plants. In regions of high winds, growth is retarded and vegetation is stunted by desiccation of exposed shoots. Near the coast and on ridges and in other open situations trees are often asymmetrical in growth due mainly to death from excessive loss of water vapour (actual drying-up) of many or all of the shoots on the windward side. Trees then look as if they have been blown or bent over towards the leeward side but the main cause of the asymmetry is inequality of shoot development on the two sides. Cold winds are among the limiting factors to tree growth at higher altitudes and higher latitudes through their desiccating action. Near the sea, salt spray is carried by strong winds even some miles inland. This may occasionally intensify the injurious effects of the wind but it is doubtful if salt spray itself does much damage.

Edaphic or Soil Factors

The vast majority of plants, except the Bacteria and Algae, are rooted in or attached to a substratum. From this is obtained, in the form of watery solutions, all the solid food materials and water. The most important substratum is soil, while rocks, walls, tree-trunks, and so on have at most a limited and specialized flora. There are also some habitats, such as recently formed sand-dunes, in which the substratum is more accurately described as " soil-to-be " than soil proper. This introduces us at once to the important concept of soil change. The particles of mineral matter, derived directly or indirectly from the disintegration and decomposition of rocks, which generally make up the original mass of a soil, undergo a profound and usually long-continued series of additions and extractions before a mature soil is made. The cause of this evolution depends on the climate, the nature of the parent rock or rocks, the vegetation, and the water supply. Chemical, physical, and biotic factors are all involved.

The realization that climate and vegetation play a great part in soil formation has been associated with a study of soils in vertical section. A face view of soil, such as can be seen in the uppermost part of a quarry or cutting or in a fairly deep trench (often made for the purpose), is termed a " profile " and examination of profiles shows that soils are layered or stratified and the different strata are termed " horizons." The number, arrangement, and detailed composition of these horizons determine the soil type. It must be remembered that ploughing of arable fields and deep digging (" trenching ") in gardens disturb the soil profile as much as manuring and fertilizing change the composition of the soil and we are concerned here only with (more or less) natural soils.

World groups of soil depend primarily on climate. In the British Isles two factors are of outstanding importance in a consideration of soils. Firstly, our climate is not extreme and, relative to temperature, is wet and humid with rain at all seasons of the year. Water, therefore, drains downwards through the soil to springs, streams, rivers, and, ultimately, the sea. Leaching-out of soluble salts is thus the rule. There is not great evaporation from the soil surface and accumulation of salts in the uppermost horizons, but rather the reverse. In fact, there is a general tendency towards increasing soil acidity. Secondly, the surface of the British Isles was much affected by the Pleistocene Ice Age. This not only directly modified the flora and vegetation but the masses of inland ice and glaciers removed soil, scoured the rocks, and in places deposited churned up masses of disintegrated products. Geologically speaking this changing of the surface is of very recent date and the result is that in the British Isles the nature of the parent rock (which may be anything from hard solid granite or schist to London Clay, Bagshot Sands, or Boulder Clay) plays a prominent part in the actual nature of the soil, which is often thin and immature when compared with soils in regions which have not been subjected to a recent geological catastrophe. There is still much to be said for a study of soils on the basis of geological stratigraphy within the British Isles. Thus in the south of England the soils on the Lower Greensand, Chalk, and London Clay are very different one from another, though developing under essentially one type of climate. It must always be remembered that possibly more types of rock (in the broad geological sense) reach to soil level in the British Isles than in any other region of equal size in the world and in many parts the outcrops alternate within very small areas.

The nature of the soil, in its influence on plant life, depends on many factors amongst which must be noted : the size of the mineral particles, ranging from coarse gravels to fine clay and silt, largely determining drainage and aeration ; the chemical composition of these particles, determining the amount of mineral food present and available for plants and the quantity of calcium in a form capable of neutralizing acidity of the soil solution ; and the quantity of matter of organic origin (humus) which plays a large part in determining both the physical nature of the soil and the availability of food materials.

There are three main " climatic " types of soil in the British Isles : " brown earths," " podsols," and " moss-peats." The brown earths occur in much of southern Britain, the midlands, and eastern Ireland. From the upper (or *A*) horizon, which may be subdivided, calcium carbonate (lime) has been leached and the soil is brown or reddish brown from hydrated ferric oxide derived from the partial decomposition of complex iron-containing minerals. As such soils are naturally formed under deciduous forest there is normally a good deal of mild humus (broken-down litter) present. Below the *A* horizon a second (or *B*) horizon is formed in which substances carried down from the upper horizon are re-deposited and this horizon is typically richer in salt content and less acid than the one above it. Below *B* there is often a " gley " horizon (*G*), grey to grey-green in colour, blotched with red-brown and due to the presence of water held in the parent rock (*C* horizon) below. Podsols occur especially in the north and west of the British Isles. They have an *A* horizon which is not only strongly leached of soluble salts but is white or grey in colour (" bleached ")

PLATE 13

F. Ballard

SAMPHIRE, *Crithmum maritimum* (Umbelliferae). Saltern Cove, Devonshire. August 1946

S. C. Porter

SEA ASTER, *Aster tripolium* (Compositae). Freston, Ipswich, Suffolk. August 1944

and tends to be markedly acid in reaction. The *B* horizon, in which the substances dissolved from *A* tend to be re-deposited, is usually sharply marked and often shows two distinct layers ; in the upper humus substances, and in the lower iron salts, are precipitated. In this *B* horizon " moorpan " or " hard pan " may be formed as a very hard cemented layer sometimes sufficiently thick and strong to prevent the downward penetration of tree roots. Coarse sandy soils may form podsols even under a brown-earth climate, as locally in southern England. Moss-peats are extremely acid and water-logged and are composed mainly of partially decomposed plant remains, especially those of bog moss (*Sphagnum*), sedges, and members of the heath family.

Other types of soil which are generally considered to depend on more local conditions, as on special water or subsoil relations, are meadow soil, fen peat, raised bog, and rendzina. Meadow soils occur in flat ground and have a water table which, though fluctuating, is never far below the surface. They are typically developed in river valleys and are covered with hay meadows or with lush pastures. Fen peat is formed in water-logged areas in which humus accumulates but is maintained in an alkaline condition through the in-draining of water rich in bases, mainly calcium. Raised bogs are acid peat accumulations formed in local basins. Rendzinas are soils derived from limestones, in this country mainly from Chalk, the Jurassic oolites, and Permian magnesian limestones. The soil is usually very shallow, alkaline in reaction, and greyish-white to dark or reddish-brown in colour. The whole profile has free calcium carbonate and leaching is slight or absent. There is no *B* horizon and the soil is more or less permanently immature.

In the study of local soil types and their relation to the plant cover the following information should be obtained : the nature of the parent rock (sandstone, clay, limestone, etc.) and its geological age ; the type of profile it shows, including the number and composition of the principal horizons and their subdivisions if any ; the seasonal variations in the level of the soil water (water table) ; and the degree of acidity or alkalinity of layers to which roots penetrate. This last is conveniently recorded by the use of chemical indicators which change colour according to the acidity, neutrality, or alkalinity of the solution to which they are added. The symbol " pH " is most often used to designate soil reaction. pH 7 is taken as neutral and a figure below 7 indicates acidity and above 7 alkalinity. Thus pH 3·5 shows an extremely acid soil, pH 8·6 a definitely alkaline soil.

Soil is not mere dirt—a dead inert mass of materials. Not only has it a chemical and physical structure and not only does it change from raw material to finished product, it also teems with living organisms and is made or modified in a very large degree by their activities. The soil flora and fauna could equally well be classified under either biotic or edaphic factors. Earthworms are referred to below under the former. Here mention may be made of the enormous population of Bacteria, Algae, Fungi, Protozoa, and insects which live, feed, multiply, and die in the upper layers of the soil. The production of humus from the breaking-down of the dead bodies of plants and animals depends on their activities. Humus alters the chemical and physical structure of the soil—indeed, changes its very nature. Further, there are soil Bacteria which fix the free nitrogen of the atmosphere into soluble nitrates which are available for higher plants, while, conversely, there are other Bacteria which break down nitrogenous salts. Soil animals and plants are responsible for a most important part of the complex of reactions which link together the climate, the soil, and all living organisms.

The water content of the soil is of paramount importance to plants. The amount and availability of water depend on precipitation and evaporation (and to this extent on climate) but also on the local topography and the nature of the soil itself. When drainage is impeded the soil becomes water-logged and this may occur either through an impermeable layer of rock, such as a band of clay or a ridge of hard rock, or because of the nature of the subsoil itself. On the other hand, the drainage may be good and the soil very pervious as in deep accumulations of loose sand. A spell of dry weather then rapidly dries out the superficial layers of the soil and the permanent water table may be lowered below root level. Plants living in water are termed hydrophytes (or aquatics), those in marshy places hygrophytes, those in habitats of medium water supply mesophytes, and those in dry habitats xerophytes. Studies relating to these will be discussed later.

Biotic Factors

Biotic factors, as understood in plant ecology, involve the action of animals and plants on plant life. From some standpoints all the organisms living in one place and mutually acting upon one another form a biotic community or unit. It is, however, often convenient to regard separately as a biotic factor any collection or group of animals or plants having marked effects upon the flora or vegetation. A major difficulty in the study of biotic factors is to obtain exact measurements of their effects. In dealing with climatic factors we have the physical instruments, and often the data, of the meteorologist, and for soil factors the methods and data of the soil chemist and soil physicist. In analysing biotic factors the biologist has to invent his own methods and standards, and since biotic factors are much more diversified than those of climate or soil, and much more difficult to control experimentally, methods of study must be correspondingly varied. Partly for these reasons, biotic factors have not yet received nearly so much attention as they deserve and constitute a wide and fascinating field awaiting more detailed investigation. Here only a brief survey will be attempted, such as may serve to indicate the scope of the problems involved. For convenience, the subject will be given the headings : animals, man, plants.

Animals. The most conspicuous activities of animals *vis-à-vis* plants are of a destructive nature. From the seed and seedling stages till the reproductive phase is again reached, plants are subjected to attack and often destruction by animals. That " all flesh is grass and all fish diatom " can be viewed from the botanical as well as from the zoological or economic standpoint. To the animal the plant is food essential to life, to the plant the animal is often a menace to the life of the individual or the species.

In part because of their size and longevity, in part because of the considerable food reserve in their seeds, some of our trees seem to have an unusually large number of enemies. Thus the oak is attacked from the acorn stage till that of old age. A large number of animals eat the acorns, the principal offenders being rabbits, mice, voles, pheasants, wood-pigeons, and jays. In the seedling stage mice and rabbits are responsible for the greatest number of deaths. In the adult stage, the oak provides food and environment for more species of insects and other invertebrates than does any other British plant. Many of these cause relatively little or no damage. Some cause the leaves or shoots to form galls and again there are more kinds of insect-stimulated galls produced by the oak than by any other British tree. Serious damage is caused by defoliation of trees by caterpillars, of which those of the green oak moth (*Tortrix viridana*) can strip a tree of its leaves in a few weeks. Mice and wood-pigeons are largely responsible for the spasmodic regeneration of beechwoods by eating enormous numbers of beech nuts and only leaving sufficient for successful establishment of seedlings in heavy mast years. Mice and rabbits also eat many of the young seedlings.

Wherever rabbits are numerous they eat all edible vegetation, the herbs down to a closely grazed turf and low shrubs like heather to rounded hummocks. Seedlings of trees and shrubs are eaten out and there is general degeneration towards a poor kind of grass heath. Sometimes a zonation can be observed around a burrow or warren. The rabbit, it should be noted, is not native to the British Isles but was introduced by man, probably in Norman times in the eleventh or twelfth century. Of other rodents, the long-tailed field mouse (*Apodemus sylvaticus*) is a seed-eater and is largely responsible for the destruction of the acorns and beechnuts already mentioned ; the bank vole (*Clethriomomys glareolus*) eats fruits, seeds, bulbs, tubers, etc. ; and the common vole (*Microtus agrestis*) is primarily a grass-eater. The common vole can alter the composition of grassland. One striking effect of a heavy vole population is the great relative increase in mosses which are not eaten at all and develop exceedingly when taller vegetation is gnawed down.

Squirrels, and nowadays in particular the introduced grey squirrel, cause considerable injury to trees by eating or biting off shoots and ringing bark. The influence of moles and ants on plant life in this country appears not to have been investigated in much detail. The flora on ant heaps is often very different from that of the immediate surroundings.

Damage caused by insect pests is usually less obvious in the wild flora than amongst cultivated crops. In the latter there are numerous, often crowded, individuals growing as pure populations under unnatural conditions. An insect pest may sweep through the crop the more rapidly because natural enemies to the insects may be absent, or present only in small numbers. There are, however, occasional striking examples of insect attack amongst native herbs. Thus the caterpillar of the cinnabar moth (*Hypocrita jacobaeae*) sometimes strips all the foliage from every available plant of ragwort (*Senecio jacobaea*) and even spreads on to other species of *Senecio*. A number of insects eat out the seeds from fruits or infructescences, as in knapweeds (*Centaurea* spp.) and other Compositae.

Cyril Newberry

a. Hairy Willowherb, *Epilobium hirsutum* (Onagraceae)
Duffield, Derbyshire. August 1943

James Fisher

b. Butterwort, *Pinguicula vulgaris* (Lentibulariaceae)
Achiltibuie, Coigach, W. Ross. May 1945

Stuart Smith

BUTTERBUR. *Petasites hybridus* (Compositae). Bank of R. Mersey near Gatley, Cheshire. March 1944

Earthworms influence plant life through their action on and in the soil. They act as cultivators, loosening and mulching the soil and facilitating aeration and drainage by their burrows. They are rich in nitrogen and their dead bodies decompose rapidly thus furnishing a certain amount of plant food to the soil. They live in slightly acid to neutral soils and favour those with high water and high organic content. Wormcasts commonly contain a higher proportion of carbonates than the surface soil and are usually less acid (or less alkaline) than the soil from which they are derived. They thus tend to obscure soil gradients and sometimes to retard such changes as in natural soils tend towards increasing acidity. Other biotic factors acting on plant life through the soil are referred to above under the heading of edaphic or soil factors.

Animals (including birds) are important agents of fruit and seed dispersal and insects in particular of flower pollination. Examples of these biotic activities are discussed in a later chapter.

Man. Of all biotic agents acting on plant life man is now the most important in our country and in the world as a whole. His influence has been two-fold—destructive and constructive. As we have seen already, the greater part of the surface of Great Britain should normally be forest-covered and was so before the advent of modern man. Forest destruction has been so great that there are now very few woods or forests in this country which are primitive. The drainage of marshes and regulation of river courses has involved the change of much marshland into pasture or arable fields. The pasturing of cattle and sheep, domestic animals whose activities must come under this heading, is in many parts of the country in enclosed fields, but on some parts of the Downs and in the Welsh, Scottish, and some north English hills there is hill or mountain grazing approximating to a range character. This has undoubtedly affected the vegetation and the change from cattle (or mixed cattle and sheep) to sheep grazing has been suggested as one cause of the great spread of bracken, particularly in Wales and Scotland.

While man has greatly modified vegetation, and thus the landscape, it is doubtful if more than a very few species have been entirely lost to our flora through man's influence. On the other hand, a very considerable number of species have been introduced by man, mostly unintentionally, and have become so naturalized as to be now important constituents of our flora. Sycamore and horse chestnut among trees and the common rhododendron (*Rhododendron ponticum*) among shrubs are certainly not native. There are hosts of ruderals and weeds that have immigrated with grain, wool, and other commercial products. It is highly probable that a large proportion of the weeds of arable land have been introduced with seeds of cultivated plants from abroad. This introduction still goes on and it is interesting to record that during the Second Great War weeds were introduced from America with seeds of carrot and other crop plants. It remains to be seen whether any of these will establish themselves here. Of introduced water plants the history of the Canadian waterweed (*Elodea canadensis*) has been the most striking. It was accidentally introduced from North America and was first found in the British Isles in 1836 at Waringstown, County Down, Ireland. On 3 August 1842 it was found in the lake of Dunse Castle, in Berwickshire, and by 1850 it had spread to many rivers, canals, ponds, lakes, and reservoirs in Great Britain, and in places had become a serious nuisance to navigation and drainage—so much so in Lincolnshire, that in 1852 a Mr. Rawlinson was sent by the Government to advise as to clearing the dykes in the fens. Dredging failed to eradicate it, but left alone it tended locally to diminish and even to die out. It is probable that the plant is not so abundant now in the country as a whole as it was from 1850 to 1880. It is generally stated that the species is dioecious, that only the female plant was introduced into the British Isles, and that its rapid spread was entirely due to vegetative multiplication—propagation by broken shoots and by winter brood-buds. The Canadian waterweed has spread widely through Central Europe and is still spreading in south-eastern Europe and may, by now, have reached Asia Minor.

The Action of Plants on Plants

Plant competition. This is a very large subject and examples of it have to be given in subsequent chapters since it enters into consideration of the composition and development of plant communities, into discussions on adaptation and natural selection, and into many aspects of plant evolution. Here we need only note the importance to these and other biological subjects of suppression of one species by another through quickness and density of growth, greater size and shading, root competition, heavy seed production, peculiarities of seed germination, successful vegetative multiplication, utilization of food materials and water, seasonal behaviour, and production of toxic substances or harmful conditions at least from decay of old and shed parts and perhaps also from definite excretions.

Parasitism. This involves direct action of plant on plant. It is of interest to remember that Bacteria are the commonest animal parasites, Fungi the commonest plant parasites, though there are some animal diseases caused by Fungi and some bacterial diseases of plants. Susceptibility, resistance, immunity, vary from species to species, from variety to variety, and even from physiological race to physiological race. A great deal is known regarding fungal parasitism of cultivated crops but much less regarding that in wild nature. Little has been done on systematic lines to determine how far fungal diseases limit range or abundance, are responsible for deaths, and act in natural selection in our wild British flora. Parasites (like broomrape and dodder) or half-parasites (like mistletoe and eyebright) which belong to the flowering plants are of less importance than fungal parasites, but their study opens up many problems (see pp. 157 *sq.*).

CHAPTER 9

PLANT COMMUNITIES, THEIR DEVELOPMENT AND MODIFICATIONS

THAT PLANTS are naturally grouped together into communities is generally recognized, as witness such common vernacular names as forest, marsh, heath, and moor. The ecologist studies the composition, structure, development, and modifications of such communities and relates them to the environmental factors. In the course of this study some general principles have emerged and it will be the purpose of this chapter to outline the characteristics of the more widespread or otherwise important communities of British plants and to illustrate these principles by reference to them. Though we have dealt first (in the last chapter) with the habitat factors, classification of the plant communities will now be based primarily on the vegetation itself and the habitat factors considered as selective, limiting, or modifying causes of the composition and distribution of the communities.

In describing plant communities we may use all characters shown by the plants as species (or other taxonomic units), as individuals, or as associates. Obviously to describe and to understand the build-up of a plant community its component species must be determined. Ecology cannot precede taxonomy (the study of plant classification). The student must, therefore, familiarize himself with the kinds of plants present in the community to be studied and know not only their names but the characters by which they can be recognized, preferably at all stages of growth from seedlings upwards. Actually it will be found that, given a sound taxonomic training to start with, ecology and taxonomy can be studied together to the reciprocal advantage of each discipline. Quite obviously research on single species has useful contributions to make to the study of plant communities. Even the behaviour of individual plants must be noted and related to conditions of the immediate environment or microhabitat. Such observations throw light on the plasticity of plants and on the factors which control and limit plant behaviour. In this chapter, however, we are concerned with plants grouped together as numerous individuals which are rarely of one species (forming pure communities) and most often of many species (forming mixed communities). In studying these communities the following features, amongst others, must be recorded : the numbers (absolute or relative) of the components, their spacing on the ground, their height and spread, the layering of their shoot and leaf systems, their underground systems whether structurally roots or stems, their seasonal behaviour, their pollination and seed-dispersal mechanisms, the germination of their seeds and the establishment of their seedlings.

Anyone who has watched a piece of vegetation over a period of years knows that it is constantly changing. Changes can be conveniently classified as seasonal, irregular, successional, or geological. Seasonal changes are periodically recurrent and are closely linked with environmental, and especially climatic, factors. Leaf-fall, bud-bursting, flowering, and fruiting are examples. Irregular changes are effected by storms, diseases, pests, and so on. Occasionally they are conspicuous in their effects and then pass over into or initiate more permanent (successional) changes, but most often they result in minor and quickly obliterated changes of quite local effect in the community as a whole —a tree is blown down, a branch decays, a grass tuft is uprooted or eaten down, but the general balance within the community is maintained. These small, irregularly but constantly occurring changes deserve much more study than they have so far received. Geological changes are those occurring over long periods connected with changes of earth history such as secular changes of climate. They cannot be observed directly and examples concerning British plant life have been discussed in earlier chapters. This leaves what are known as successional changes, or simply plant succession. An area of bare ground quickly or slowly becomes invaded by plants. If left undisturbed, a definite sequence of changes occurs in which species and communities replace one another in a succession which is controlled by determinable causes. It is thus possible to formulate general laws of succession. A primary succession is one starting from bare ground which has not previously had vegetation growing on it or, more precisely, which does not contain any disseminules (seeds, bulbs, rhizomes, spores, or anything that can grow up into a mature plant). Any vegetation which appears is due to immigration. A secondary succession arises on an area which has previously carried vegetation and on which the succession does not start from scratch—disseminules of some kind and usually a soil, as distinct from mere terrain or substratum, are already present. Indeed, a secondary succession may start or restart at any stage from complete destruction of all growing plants, as after a severe fire, to the cutting out of mature timber only from a forest. Two important principles of succession are :

1. That at any stage the vegetation is itself helping to change the habitat in such a way as to make it less and less optimal for the existing vegetation and more and more suitable for the vegetation of the next stage which gradually replaces it ; and

2. that such changes eventually reach a limit in what is known as the climax stage (or, simply, climax) and this is most often forest in the British Isles except at higher altitudes and higher latitudes. An actual succession is termed a sere, and usually within the climatic conditions of the British Isles a sere progresses from a more xerophytic or a more hydrophytic stage to a mesophytic climax. In simpler terms, vegetation, in our climate, usually changes naturally from communities living either under very dry or very wet conditions to a final stage in which the vegetation has and maintains surroundings that are not extreme. There are various ways in which a sere can be halted more or less permanently at a given stage or deflected from its normal succession, and naturally or artificially it may at any stage be, so to speak, amputated and caused to rejuvenate in a secondary succession.

The purposes to be served by a classification are often numerous and priority has sometimes to be given to one purpose over others. Several classifications are possible for British plant communities. A classification based on succession has much to recommend it but would be unduly complicated for our present purpose and would involve a good deal of unfamiliar terminology—some would say unnecessary scientific jargon. In the present state of our knowledge of vegetation it would also be very incomplete. A simple classification based primarily on general physiognomy has therefore been prepared in such a way as to include all the major types of community and at the same time to allow the interpolation of successional data and correlations with habitat factors. The scheme is as follows :

Natural and semi-natural communities

Maritime submerged vegetation
Coastal communities : salt marshes, sand stretches and sand dunes, shingle beaches, rocks and cliffs
Aquatic communities
Marshes
Bogs

Grasslands
Heaths and moors and brushwoods
Woodlands
Mountain communities

Artificial communities

Arable farmland
Artificial grasslands
Gardens and orchards
Plantations
Ruderal communities

Natural and Semi-natural Communities

Maritime submerged vegetation. In the sea itself around our coasts marine Algae (seaweeds) constitute the sole vegetation with the exception of the eel-grasses or sea-grasses (two or three species of *Zostera*). The eel-grasses occur on muddy flats, usually just above to just below low spring-tide mark. Sometimes eel-grasses occur in such quantities as to form veritable submarine meadows. A few years ago a disease attacked at least one of the species and caused considerable reduction in the extent of the meadows. The species of *Zostera* are flowering and seed-bearing plants with many peculiarities of structure adapting them to the unusual conditions of their habitat. In particular, the flowers open submerged and produce rather long and slender stigmata (portions of the " female " parts of flowers receptive of the pollen) while the pollen grains are thread-like and of the same specific gravity as the sea water, so that when discharged they tend to float at the level of discharge, which is approximately that of the stigmata, and neither to sink to the bottom nor rise to the surface.

The seaweeds between and below the tide marks are a special study. It is of interest here to note that they are most abundant on rocky shores and usually show zonation which varies with the nature of the rock and exposure to wave action, while the zonation itself in any one place is largely controlled by the time of exposure (at low tide) and the quantity and quality of light which penetrates to them. A broad general sequence, with increasing depth, of green, brown, and red seaweeds can sometimes be observed.

Coastal Communities

The extent of our coasts and the varied nature of their topography and geological composition are reflected in the excellent development of plant communities which are often diverse within quite short distances. These communities are all exposed to sea winds and some to periodic immersion in sea water or to the effects of sea spray. Their structure and successional stages, however, are of exceptional interest because in many places they have been little or not at all interfered with by man. Moreover they are very different one from another in their composition and development.

Salt marshes. These are composed of halophytes or plants able to withstand high salt concentration around their roots and submersion in sea water. They are developed on low protected shores and in estuaries where there is a mud substratum, and are periodically immersed by the tide. Different levels of the salt marsh are covered by sea water for very different periods, from the highest parts only reached by high spring tides to those covered by high water of neap tides. A general zonation (though in any one locality one or more zones may be absent or indistinct) is from lower levels upwards as follows:

1. Samphire or glasswort marsh, dominated by *Salicornia herbacea.*
2. Rice-grass marsh, with *Spartina townsendii*, of local occurrence in the south of England.
3. Sea grass marsh, of *Glyceria maritima.*
4. Sea aster marsh, with *Aster tripolium* (Pl. 14, p. 79) the most conspicuous species.
5. Sea lavender marsh, with *Limonium vulgare* dominant or co-dominant with *Glyceria maritima.*
6. Thrift marsh, dominated by *Armeria maritima.*
7. Sea purslane marsh, with the greyish white sub-shrub *Obione portulacoides* the most important plant.
8. Seablite marsh, of *Suaeda fruticosa*, occurring locally on the south and east coasts of England and often marking the boundary between salt marsh and shingle beach.
9. Red fescue marsh, of *Festuca rubra*, occurring especially where the mud becomes mixed with sand and above the general salt-marsh level.
10. Sea rush marsh with *Juncus maritimus* dominant and representing transition from salt marsh to dry land.

F. Ballard

HEATHER, *Calluna vulgaris* (Ericaceae)
Bladon Heath, near Woodstock, Oxfordshire. August 1944

John Markham

CHICKWEED WINTER-GREEN, *Trientalis europaea* (Primulaceae)
Rothiemurchus, Inverness-shire. June 1946

To a considerable extent these zones represent stages in succession. Deposition of silt brought in by the tides and the growth and decay of vegetation raise the general level and thus reduce the time of submergence and the amount of scour and improve the drainage. At the same time the salt marsh is continually being broken down. Sometimes there is marked erosion at a seaward edge, nearly always there is erosion at the margins and ends of the numerous creeks up which the tide swirls as it flows and into which it drains off the salt marsh as it ebbs. "Pans" or stagnant pools of salt water are of frequent occurrence on salt marshes. They are continually being formed or enlarged and as continually being silted up and covered again by vegetation, though some may persist for long periods and even open up into channels. Pans are formed in various ways and need further investigation. In general terms, the salt marsh is constantly undergoing change: in one place it is being built up, in another pulled down. If artificially enclosed by dykes, it speedily changes its character and may become useful agricultural land, but it is doubtful how far salt marshes on our British coasts, with their strong tidal action, can progress naturally into other communities unless topographical changes occur to the coast-line.

The extreme halophytic nature of the salt-marsh flora is an exceptionally clear example of natural selection by environmental factors on an ecological scale. The flora is limited to those plants which can withstand high salinity and a considerable degree of fluctuation in this salinity, for parts of the salt marsh may have sea water concentrated to saturation in a dry spell or diluted in a rainy spell. Moreover, the typical salt-marsh plants occur only in this community though some of them can be cultivated successfully in ordinary garden soil in the absence of competition. Besides the salt-marsh plants mentioned above as characterizing zones there are others such as sea plantain (*Plantago maritima*), sea arrow-grass (*Triglochin maritimum*), sea milkwort (*Glaux maritima*), seablite (*Suaeda maritima*), sea spurreys (*Spergularia* spp.), and scurvy grasses (*Cochlearia* spp.), all of which are halophytes, either obligatory or facultative. In addition certain Algae are important members of some salt marshes and some are present as remarkable variants of the species.

Sand stretches and sand-dunes. These have a vegetation which, except near the seaward margin, is not halophytic but shows some characters typical of plants of dry places. Nearest the shore where sea spray and even very high tides may affect any plants occurring there is little vegetation except sometimes widely spaced plants of sea rocket (*Cakile maritima*), saltwort (*Salsola kali*), or, more rarely, sea kale (*Crambe maritima*). Foreshore vegetation still requires detailed investigation. Most areas of sand of any considerable size near the coast above high-tide mark are piled up by on-shore winds into sand-dunes which usually form ridges parallel with the coast. Such sand-dunes are mobile until they are clothed with and fixed by vegetation. The commonest pioneer colonist on sand-dunes is the marram-grass (*Ammophila arenaria*, sometimes called *Psamma arenaria*). This is a tall grass which is not only structurally suited to the environmental conditions but reacts to them and in time changes them. It has an extremely extensive under-sand system of rhizomes (elongated stems growing underground) and roots which help to bind and hold down the sand particles, but it grows through sand blown on top of it and flourishes best when more sand is frequently added. It is used a great deal in the artificial stabilization of sand-dunes. The sea couchgrass (*Agropyron junceum*) and lymegrass (*Elymus arenarius*) are other species of sand-dune grasses. The marram-grass is often the only species occurring in any abundance in the early stages of sand-dune fixation but later many other plants immigrate from the land vegetation. It is the semi-mobile, loose, highly permeable substratum with little humus which, in the early stages of stabilization, limits the floristic composition, and not salinity or competition between species. Three groups of plants are characteristic of the middle stages of dune fixation: 1. annuals, or rather ephemerals, which flower and fruit very quickly in early spring and include mouse-eared chickweeds (*Cerastium* spp.), whitlow grass (*Erophila verna*), wall saxifrage (*Saxifraga tridactylites*), and early scorpion-grass (*Myosotis collina*); 2. perennial or biennial herbs as creeping thistle (*Cirsium arvense*), ragwort (*Senecio jacobaea*), stonecrop (*Sedum acre*), sand sedge (*Carex arenaria*), hound's-tongue (*Cynoglossum officinale*), and dandelions (*Taraxacum*); 3. mosses, liverworts, and lichens. Mosses and lichens are particularly abundant and by their growth close to the soil, often as compact carpets, afford considerable protection against wind action, and as they decay they add humus to the soil, increasing both plant food for higher plants and its water-retaining capacity. In the later stages of development the vegetation consists of a host of ordinary inland species—herbs, shrubs, and eventually trees in so far as these obtain some protection from winds. There is a general tendency for sand-dunes to start with an alkaline or neutral substratum which, however, becomes more acid as the calcareous materials (such as comminuted shells) are neutralized and acid humus accumulates. Later stages of the vegetation often resemble heath vegetation and finally heath woodland.

Between the dune-hills or ridges, especially on our west coast, there are frequently damp or wet hollows known as "slacks." The water table is here high and there may even be ponds formed. There is in and around these slacks a development of general marsh and aquatic plants. Mosses and liverworts are abundant and willows (creeping willow, *Salix repens*, and sometimes taller-growing species) may form low to fairly tall marsh brushwoods.

There are many fine sand-dune areas around the British coasts. Typical examples are those of Blakeney Point and Scolt Head in Norfolk, Braunton Burrows in N. Devon, near Southport in Lancashire, near Harlech and Llanbedr in Merionethshire, and the Culbin Sands on the Moray Firth.

Shingle beaches. Where the substratum is composed of rounded pebbles, usually around the English coasts of flint, the vegetation is almost or quite absent on the seaward slope and crest, and open, except locally, on much of the landward slope owing to mobility of the shingle in on-shore gales at high tides. Rather unexpectedly there is plenty of water in shingle beaches and this is commonly fresh. Shingle plants do not suffer from drought and are not halophytes. The amount of humus present varies and, above high-storm mark, the selective factors of the habitat are mobility of the pebbles and the exposed situation of most shingle beaches. Characteristic plants of the shingle beaches include the shrubby seablite (*Suaeda fruticosa*), noteworthy in Norfolk and on the Chesil Beach, the yellow horned poppy (*Glaucium flavum*), a dock (*Rumex crispus* var. *trigranulatus*), the sea campion (*Silene maritima*) (Pl. 38b, p. 179), sea couchgrasses (*Agropyron pungens* and *A. junceum*), red fescue (*Festuca rubra*), sowthistles (*Sonchus* spp.), sea purslane (*Honckenya peploides*), and stonecrop (*Sedum acre*). Where the shingle has been stable for some long time lichens become very numerous. Transition to salt-marshes is usually very abrupt but mixtures of sand and shingle, or one overlying the other, are not infrequent. There are excellent examples of shingle beaches around the British coasts and the following may be named: Pevensey in Sussex (fringing beach in contact with the land), Hurst Castle in Hants and Blakeney in Norfolk (shingle spits carried out beyond the shore), Dungeness (apposition shingle beach formed by a succession of more or less parallel banks), and Chesil Bank (a shingle bar enclosing a lagoon).

Rocks, banks, cliffs. Rocks, banks, and cliffs near the sea, when out of reach of the tides but much exposed to sea spray, have a more or less halophytic and usually open vegetation. Samphire (*Crithmum maritimum*) (Pl. 13, p. 78) is very typical of such habitats and other species found under such conditions are : sea beet (*Beta maritima*), sea campion, sea lavender, and sea spurreys. Above or beyond the constant effects of sea spray the vegetation becomes more mixed, more closed, and its floristic composition more of an inland type, but the plants usually remain dwarf except in well-protected pockets. A number of plants which are rare or local in Britain occur on sea cliffs, and these include the wild cabbage (*Brassica oleracea*), queen stock (*Matthiola incana*), and some sea lavenders (*Limonium* spp.). The vegetation of sea cliffs varies greatly and needs further detailed and comparative ecological study. A range behaviour which remains unexplained is the limitation of certain species usually to within a few miles of the sea, as happens with alexanders (*Smyrnium olusatrum*), fennel (*Foeniculum vulgare*), and slender thistle (*Carduus tenuiflorus*).

Aquatic Communities

There is, as we have seen in an earlier chapter, very strong evidence that plant life originated in water and later invaded the dry land. This invasion, however, occurred long before the appearance of the flowering plants—indeed flowers and seeds are very definitely advanced structural adaptations to terrestrial life. It follows that while such water plants as some members of the Algae may be primitive, aquatic flowering plants have been derived from land ancestors. There are certain advantages, particularly for vegetative stages, of life in water. Most prominent is an abundant supply of water itself with no danger of drought or of transpiration exceeding water-intake. One disadvantage is the difficulty of adequate gaseous exchange for respiration and photosynthesis. Oxygen supply in particular may be low. Anchorage and food supply are sometimes limiting factors. Freely floating plants must obtain all their food from the surrounding water ; attached aquatics have been shown to draw at least much of their food from the mud and silt of the bottom. Both suitable rooting substratum and dissolved

Brian Perkins

COWBERRY, *Vaccinium vitis-idaea* (Vacciniaceae). Ben Vrackie, Perthshire. June 1946

John Markham

a. COMMON OAK, *Quercus robur* (Fagaceae)

John Markham

b. DURMAST OAK, *Quercus petraea* (Fagaceae)

food may be largely absent from lakes in rocky basins or in high mountain tarns and the aquatic vegetation is then very poor. The greatest risks, however, are concerned with flowering and pollination and these have to be minimized by devices of structure and behaviour. The marked peculiarities of the environment as contrasted with terrestrial habitats are correlated with very obvious adaptations shown by the plants themselves (see p. 148). Some aquatics flower and fruit submerged, but the majority have their flowers at or above the surface.

Succession, on a smaller or a larger scale, can easily be studied in many aquatic habitats. The succession is a hydrosere—i.e. it starts from deep open water and marsh stages and eventually attains to dry land. Silt and dead organic remains build up the bottom and make the habitat less suitable for deeper water plants and more suitable for shallower water and finally terrestrial species. Around many lakes and ponds there is a marked zonation of the vegetation and, within limits, the zones illustrate the successional stages.

Ponds and lakes. Small ponds when stagnant, shaded, and with excess of decaying organic matter have little plant life other than Bacteria and blue-green Algae. Many lowland ponds, however, receive water which has drained over soft strata and is rich in silt and dissolved food materials. Such ponds have often a lush aquatic vegetation and a flora which may vary greatly from pond to pond. Unless periodically cleaned out they quickly pass through successional stages and become small marshy patches and eventually merge with the surrounding community. Submerged constituents may include: many species of green Algae, species of stoneworts (*Chara* and *Nitella*—often classified as Algae), pondweeds (*Potamogeton* spp.), water-milfoils (*Myriophyllum* spp.), hornworts (*Ceratophyllum* spp.), shoreweed (*Littorella uniflora*), Canadian waterweed (*Elodea canadensis*), and bladderwort (*Utricularia vulgaris*). Plants largely submerged but with the leaves, or some of them, floating at the surface may be represented by certain water buttercups (*Ranunculus peltatus*, etc.), frogbit (*Hydrocharis morsus-ranae*), yellow water-lily (*Nuphar luteum*), white water-lily (*Nymphaea alba*) (Pl. VI, p. 143), broad-leaved pondweed (*Potamogeton natans*), floating persicaria (*Polygonum amphibium*), flotegrass (*Glyceria fluitans*), and star-worts (*Callitriche* spp.). Many small ponds have duckweeds (especially *Lemna minor*) floating freely on the surface. A little below the surface *Lemna trisulca* may be found. Populations of *Lemna* should be carefully examined in the summer for the very small flowers. Flowering in some duckweeds is rare to very rare, but when it occurs practically every plant in the population may flower. Some aquatic species, rooted in the mud, grow up above the water surface and produce flowers in the aerial portion. This behaviour is typical of water violet (*Hottonia palustris*), mare's-tail (*Hippuris vulgaris*), and bog bean (*Menyanthes trifoliata*). In larger ponds or lakes and sometimes along the margins of slowly flowing rivers there may be a reed-swamp developed. This is a community in some respects intermediate between aquatic and marsh communities. Frequently one species may be so dominant as to form an almost pure population. Typical constituents are bulrush (*Scirpus lacustris*), water horsetail (*Equisetum fluviatile*), reed maces (*Typha latifolia* and *T. angustifolia*), common reed (*Phragmites communis*), some sedges (*Carex* spp.), and reed meadow-grass (*Glyceria maxima*).

Aquatic communities are widely scattered throughout the British Isles, though some counties are richer than others in such suitable habitats, as ponds, lakes, rivers, and streams. The Norfolk Broads are shallow lakes with a rich aquatic vegetation whose luxuriance is partly connected with the food supply present in the water which is of alkaline reaction (pH above 7), through the drainage being over soft and largely lime-containing rocks. The Broads are mostly enclosed in fen peat. Another, very different, area is that of the English Lake District. Here are the largest of English lakes, developed in glacial lake basins some of which have very little, and others deep, silt. The water of these lakes is very pure and its "softness" contrasts with the "hardness" of the water of the Broads. A series can be arranged from the more primitive lakes with a rocky substratum to those which have a deep layer of silt. Besides the nature of the bottom, the depth of water (influencing penetration of light to aquatic vegetation), and temperature differences help to control the distribution of species and the types of vegetation in the lakes.

Rivers and streams. An important factor, present in these but absent in ponds and lakes, is the existence of a current constant in one direction. This erodes the banks and increases both the amount of oxygen in the water and the general circulation of gases and dissolved food materials. When the current is very rapid, as in some mountain streams, and especially when it reaches a maximum in waterfalls and torrents, plants have difficulty in obtaining or maintaining root-hold and the vegetation is often limited to Algae, mosses, and liverworts. In rivers with a moderate current the vegetation is denser and dominated by flowering plants. These may be arranged in a series of zones : submerged communities in the deeper parts, most open where the current is most powerful, followed by a zone of floating leaf species, and reed-swamp nearest the bank. Where the current is slow or almost absent vegetation may be very dense, zonation well marked, and successional stages passed through relatively quickly. It is, for example, surprising how quickly disused canals become blocked with masses of aquatic and, in due course, marsh plants. The seasonal aspects of aquatic vegetation, whether in ponds and lakes or in rivers, are well worth careful study.

Marshes

It is desirable to distinguish between marsh, fen, and bog. In all of these the summer water-level is near, at, or slightly above the ground surface. In a marsh the substratum has an inorganic (mineral) or muddy basis ; in a fen the substratum is organic (peaty) but the reaction of the water is generally neutral to alkaline ; in a bog there is often deep acid peat. Marsh and fen are very similar in their vegetation, and are here considered under the same main heading, but contrast markedly with bog in many ways.

Typical marshes occur on low river banks, flood plains, and lake shores. Marsh vegetation is commonly zoned and the zones often form transitions from the aquatic vegetation of open water to the dry-land vegetation of grassland, brushwood, or woodland. The reed-swamp, already mentioned, often forms, when present, an intermediate zone between aquatic and marsh communities. It grows in deeper water than ordinary marsh and is most often dominated by a single species (nearly always a monocotyledon). Typical marsh has frequently a very mixed vegetation of tall, lush-growing plants. The decay of marsh vegetation results not in peat but in black ooze or mud. The marsh flora of Britain is a very large one and only a few examples can be quoted here as of particular interest :

kingcup or marsh marigold (*Caltha palustris*)
sedges (*Carex* spp.)
rushes (*Juncus* spp.)
purple loosestrife (*Lythrum salicaria*)
water forget-me-nots (*Myosotis* spp.)
water ragwort (*Senecio aquaticus*)

hairy willowherb (*Epilobium hirsutum*) (Pl. 15a, p. 82)
marsh pennywort (*Hydrocotyle vulgaris*)
marsh docks (*Rumex* spp.)
gipsy-wort (*Lycopus europaeus*)
meadow-sweet (*Filipendula ulmaria*)
marsh orchids (*Orchis* spp.)
yellow flag (*Iris pseudacorus*)
water-mints (*Mentha* spp.)
hemlock water dropwort (*Oenanthe crocata*)
water bedstraws (*Galium* spp.)
great spearwort (*Ranunculus lingua*)
fleabane (*Pulicaria dysenterica*) (Pl. 31, p. 126)
skullcap (*Scutellaria galericulata*)
sweet flag (*Acorus calamus*)
marsh arrow-grass (*Triglochin palustre*)
valerian (*Valeriana officinalis*)
grass of Parnassus (*Parnassia palustris*) (Pl. 10a, p. 67)
butterbur (*Petasites hybridus*) (Pl. 16, p. 83)

In addition there are a considerable number of marsh mosses, especially species of *Hypnum*.

The fens of East Anglia, although now much reduced by drainage and reclamation, still form the most extensive area of marshlands in the British Isles. The drained peat and silt is very fertile and on it potatoes, sugar beet, bulbs, and fruit grow well. The vegetation of the fens still remaining (Wicken, Chippenham, etc.) has been much modified by cutting and other human activities. The list of fen species is a long one and dominance varies greatly according to the degree and kind of treatment a given area (sometimes a strip) has received and the stage of succession an area has reached. Dominants, in one place or another, include common reed, reed meadow-grass, reed-grass (*Phalaris arundinacea*), purple moor-grass (*Molinia caerulea*), and common sedge (*Cladium mariscus*). Brushwood (" carr ") passing into woodland develops naturally on fen at about the height of the winter water table. At Wicken Fen early shrub colonizers are alder buckthorn (*Frangula alnus*), grey sallow (*Salix atrocinerea*), common buckthorn (*Rhamnus cathartica*), and guelder rose (*Viburnum opulus*). In the East Norfolk fens, ash (*Fraxinus excelsior*), hairy birch (*Betula pubescens*), and alder (*Alnus glutinosa*) enter into the composition of carr, the two latter becoming dominants. It is probable that oak forest would naturally succeed fen if there were no interference by man. When fen vegetation, especially sedge, is subjected to regular cutting, a deflected succession

F. Ballard

TOOTHWORT, *Lathraea squamaria* (Orobanchaceae), parasitic on hazel (*Corylus avellana*)
Glympton Wood, near Woodstock, Oxfordshire. April 1945

PLATE 22

John Markham

Mountain Ash, *Sorbus aucuparia* (Rosaceae) by Highland river
Rothiemurchus, Inverness-shire. June 1946

results. Bushes cannot establish themselves, but as the ground becomes drier *Molinia* (locally termed " litter ") increases and in time replaces *Cladium*.

BOGS

Bogs (" mosses " of northern England and Scotland) are plant communities growing on and forming acid peat. They are most extensively developed in the north and west of the British Isles and in southern England are confined to local depressions mainly on sandy soils. In bogs, the water table is practically at or locally slightly above the surface. Various types can be distinguished. Valley bogs occur where relatively acid water stagnates in a flat-bottomed valley and bog moss (*Sphagnum*) and other plants characteristic of a wet acid substratum accumulate. There is generally sufficient drainage into and through a valley bog to maintain some supply of mineral salts and oxygen circulation and, as a result, species of rushes (*Juncus*) and sedges (*Carex*), such as never occur in extreme types of bog, may maintain themselves. A raised bog has a slightly convex surface and as a whole is raised above the immediate surroundings. It is most often developed on lake and fen deposits and its water supply depends on direct precipitation. The whole bog is extremely acid and its flora limited. A blanket bog depends on high rainfall and high atmospheric humidity and is thus more dependent on climatic conditions than are the other types of bogs. It is the commonest type in the north and west where flat blanket bogs often cover considerable areas in a manner likened to a blanket.

Bog-mosses (*Sphagnum*), of which there are numerous species varying in the details of their habitat requirements, are the most abundant plants of bogs. The absorptive and antiseptic properties of bog-mosses have led to their use in pads for dressing wounds. Other plants occurring in bogs, sometimes in sufficient numbers to give a name to the plant community, are: cotton-grass (*Eriophorum angustifolium*), white beak-rush (*Rhynchospora alba*), bog asphodel (*Narthecium ossifragum*) (Pl. VII, p. 158), black bog-rush (*Schoenus nigricans*), deer's grass (*Scirpus caespitosus*), sundews (*Drosera* spp.), butterwort (*Pinguicula vulgaris*) (Pls. VIII, p. 159, and 15b, p. 82), and sweet gale (*Myrica gale*). When the surface is somewhat drier there may be considerable development of cross-leaved heath (*Erica tetralix*), heather (*Calluna vulgaris*), and purple moor-grass.

Grasslands

Grasslands, in the broad sense, occupy a larger area in the British Isles than any other physiognomic group of plant communities. Few of them, however, are climatic climaxes or end-communities of a succession. If allowed to develop naturally, without grazing or cutting for hay, the vast majority would pass through brushwood to woodland. The British climate is undoubtedly very favourable to grassland, but it is also favourable to woodland and with natural competition woodland succeeds grassland and maintains its climax position. It follows that most of our grasslands are either artificial or are artificially maintained as such. There are all degrees of human control ranging from sowing grasslands, whether as temporary leys or as permanent swards, including lawns, to the mere cutting-out of shrubs and trees and the effects of grazing. Grazing itself may be intense or of a casual, wide-ranging type. There are marked effects of overgrazing which may include the formation of a close turf of a limited number of species or the appearance of bare patches which become occupied by unpalatable weeds. Undergrazing may be as harmful to pasture, from the human economic standpoint, as overgrazing, since tall rank-growing herbs may come to predominate and the area be also invaded by shrubs. Intensive grazing at one or more seasons and rest periods in between also modify the flora and vegetation of pastures. The kind of herbivore using the pasture is also a factor. Thus the reduction in cattle and the great increase in sheep on the hill pastures of Wales and Scotland is said to be one cause of the deterioration of these rough grazings, evident in the spread of bracken and deer grass.

Since most of our grasslands are maintained, if they have not been formed, by pasturing or cutting for hay and are not natural climaxes, the most satisfactory classification is one based on the vegetation as correlated with soils and especially with the pH of the soil water. The following arrangement covers the more important types, though there are a great many local variations and much comparative research remains to be done on the composition of grassland and the factors controlling and modifying it in the British Isles.

Basic grassland. This occurs on chalk, limestones, and other basic rocks where the soil is shallow, well drained, and alkaline in reaction. Sheep's fescue (*Festuca ovina*) and red fescue (*F. rubra*) are common grasses, especially where there is well-controlled grazing. If grazing is badly managed coarser grasses (especially *Brachypodium silvaticum* and *B. pinnatum*) may become dominant and in winter are conspicuous because of the dead straw-coloured tufts of haulms and foliage. Where grazing is absent the erect brome (*Bromus erectus*), in flower a very beautiful tall-growing grass, will oust not only other grasses but many other herbs, though, if the area be left undisturbed, it may itself be replaced by shrubs and woodland. The flora of basic grasslands is relatively rich and, especially on the English Chalk, contains a fair proportion of species of southern range. Leguminous plants and orchids are sometimes numerous.

Acidic grassland occurs on sands, some clays, metamorphic and igneous rocks of a siliceous character, and locally on other substrata where the soil reaction becomes acid. Three major types have been described. Bent-fescue grassland is dominated by species of *Agrostis* and *Festuca* and probably occupies a larger area than any other kind of semi-natural grassland in the British Isles. Where heath vegetation has been partially destroyed by cutting or burning of brushwood a modified type of bent-fescue grassland may occur with waved hair-grass (*Deschampsia flexuosa*) or soft grass (*Holcus mollis*) dominant. Mat grass communities of *Nardus stricta* occur on siliceous soils of medium dampness where acid peaty humus has accumulated. It is of little or no grazing value and is floristically poor. Purple moor-grass community with *Molinia caerulea* as the dominant grass occurs on continually wet, more or less peaty, soil. Transitions between acidic grasslands and heaths and moors are frequent. Bracken (*Pteridium aquilinum*) often spreads in bent-fescue grassland and reduces its grazing value. Gorse scrub (of *Ulex europaeus* and, in the west, *U. gallii*) is also a feature of many bent-fescue grasslands.

Neutral grasslands is a name applied to grasslands developed on soil, clays, loams, and alluvia, nearly neutral to slightly acid in reaction. Most of them are subjected to intensive grazing, mowing, or trampling. Some of the best pastures for the farmer are those with abundant perennial rye-grass (*Lolium perenne*) and wild white clover (*Trifolium repens*). Other valuable grasses are cock's-foot (*Dactylis glomerata*), timothy (*Phleum pratense* and *P. nodosum*), crested dog's-tail (*Cynosurus cristatus*), meadow-grasses (*Poa pratensis* and *P. trivialis*), meadow foxtail (*Alopecurus pratensis*), sweet vernal grass (*Anthoxanthum odoratum*), and meadow fescue (*Festuca pratensis*). Coarser grasses are Yorkshire fog (*Holcus lanatus*) and tufted hair-grass (*Deschampsia caespitosa*).

Road-verges, commons, parklands, and lawns are more or less modified grasslands and often illustrate very clearly the action of some biotic factor which maintains both a distinctive flora and a definite physiognomic kind of vegetation. Newly formed road-verges can show how very quickly early successional stages may replace one another and there is yet much to be learnt from their study.

It must be remembered not only that the term " grasslands " includes a wide range of communities but that the flora is often not by any means solely composed of grasses, though usually there are more individuals of grasses than of any other family. Leguminous species are often important and include clovers (*Trifolium* spp.), black medick (*Medicago lupulina*), bird's-foot trefoil (*Lotus corniculatus*), vetches (*Vicia* spp.), and meadow vetchling (*Lathyrus pratensis*). In chalk grasslands horse-shoe vetch (*Hippocrepis comosa*) is sometimes abundant. The Leguminosae are important in that their root-nodules harbour nitrogen-fixing Bacteria which enrich the soil by adding soluble salts containing nitrogen. In addition to grasses and legumes a host of other herbs occur in grassland. Some of these are weeds, from the farmer's grazing or hay-making standpoints, but others are extremely valuable in supplying to stock mineral and other foods which are lacking, or occur in only small quantities in grasses and legumes. This, for example, appears to be true of ribwort plantain (*Plantago lanceolata*).

Heaths and Moors

Attempts to distinguish between heaths and moors have not been very successful and here the communities on acid soil dominated usually by low shrubs, of which heather or ling (*Calluna vulgaris*) (Pl. 17, p. 90) is the commonest, are considered under one heading. Recurrent fires, high winds, grazing, or tree-cutting prevent the formation of tall brushwood or woodland. High intensity of grazing may convert heath into grassland.

The heaths of southern England occur mainly on sands and gravels with the poor soil a podsol (see p. 78). Iron-pan formation is not uncommon. Fires are frequent in dry seasons and many stages of secondary succession can be studied on heaths and even on commons near London (Ham Common, Putney Heath, Sheen Common, Oxshot Heath, Hindhead Common, etc.). These southern lowland heaths vary in structure and composition not only according to the stage of development after fire but also according to the amount of very direct human interference in the form of cutting and trampling. Heather is the most characteristic species, but others that should be mentioned are: dwarf gorse (*Ulex minor*), bilberry (*Vaccinium myrtillus*), bell-heath (*Erica cinerea*), heath bedstraw (*Galium saxatile*), lesser sheep's sorrel (*Rumex acetosella*), milkwort (*Polygala vulgaris*), and heath grasses (*Deschampsia flexuosa, Sieglingia decumbens*, etc.). Bracken and common gorse (*Ulex europaeus*) frequently invade heather areas. Heath mosses are common and characteristic and include *Hypnum cupressiforme* var. *ericetorum*, *Dicranum scoparium*, *Polytrichum* spp., and *Leucobryum glaucum*, while lichens of the genus *Cladonia* are frequent.

In constantly wet (water-logged) areas within heathland, wet heath or heath bog may develop. Purple moor-grass is a common dominant in some wet areas and is often associated with cross-leaved heath (*Erica tetralix*). There is no sharp line between such wet heathy areas and local bogs with *Sphagnum* and sundew.

In the south-west of England there are heaths, on non-calcareous rocks, characterized by having, often in considerable numbers and locally dominant, some southern species of heath (*Erica*) not occurring in the midlands or north of the country. On the Lizard Peninsula there are considerable areas with Cornish heath (*Erica vagans*) and in both Dorset and Cornwall ciliate heath (*Erica ciliaris*) forms communities.

In the north of England and in Scotland are the so-called heather moors which in many ways are similar to the southern heaths. Heather is again a usual dominant but there are a few species which occur in the northern heather moors but not (or very rarely) in the south of England. Amongst these may be mentioned the shrubs crowberry (*Empetrum nigrum*), bearberry (*Arctostaphylos uva-ursi*), cowberry (*Vaccinium vitis-idaea*) (Pl. 19, p. 94), cloudberry (*Rubus chamaemorus*), and dwarf birch (*Betula nana*) and the herbs chickweed winter-green (*Trientalis europaea*) (Pl. 18, p. 91) and winter-greens (*Pyrola* spp.). The moors of the Pennines and other parts of northern England and the grouse moors of Scotland come under this group of northern heather moors.

BRUSHWOODS

Brushwoods are communities dominated by shrubs or bushes. Most brushwoods in the British Isles occupy areas formerly covered by forest and would revert again to forest if allowed to complete their seral development. Only in a few areas, especially near the sea and on high ground, where there is exposure to frequent strong winds, is brushwood likely to represent a climatic climax. Brushwoods deserve more detailed study than they have so far received in this country and the following account should be supplemented by the reader's own observations.

Thickets or thicket scrubs are dense communities devoid of trees and formed mainly or entirely of thorny species, especially gorse (*Ulex europaeus*), sloe or blackthorn (*Prunus spinosa*) (Pl. IX, p. 174), hawthorns (*Crataegus* spp.), blackberries (*Rubus* spp.), and roses (*Rosa* spp.). Hazel (*Corylus avellana*) and dogwood (*Cornus sanguinea*) may be associated with the thorny shrubs. Rabbit grazing is sometimes a factor in stabilizing this form of brushwood.

Coppice brushwood is often the shrub layer beneath well-spaced standard trees, particularly oaks, but occasionally hazel or sweet chestnut coppice, usually cut at intervals of 7 to 12 years, occurs without standards. Osier beds of willows (*Salix* spp.) are coppice brushwoods of wet places, especially river valleys.

" Woodland scrub " is the name given to brushwood which is a seral (developmental) stage leading to some kind of woodland. It may develop on abandoned agricultural land. At first there are weeds of arable land or pasture and pasture weeds, then invasion by shrubs from hedgerows or wood margins together with marginal woodland herbs, then young trees establish themselves, and finally a woodland shade flora appears. The species involved in the different stages vary with the environmental factors and particularly with the nature of the soil. Chalk scrub, limestone scrub, and heath scrub show a number of differences in structure and composition.

Hedgerows offer many problems to the ecologist. Many of them represent, in their composition, relicts of marginal woodland brushwoods. Where they are regularly cut and trained the number of species may be small, hawthorns being the commonest and often planted constituents. Sometimes hedges are allowed to grow tall and to form shelter belts. They then harbour a mixed flora of woodland, brushwood, and grassland species whose interrelationships and reactions illustrate many ecological principles.

Woodlands

The greater part of the British Isles was formerly covered by high forest. There is also no doubt that the natural environment is suitable to tree growth, with only local exceptions. Man has not only destroyed most of our forests but has highly modified those that now exist. A considerable number of our woodlands have been planted by man on deforested sites which may for a longer or shorter interval have been arable land or pasture. When planted with exotics, such plantations are obviously artificial but when planted with native species the resultant woods may become in time practically indistinguishable from natural or semi-natural woods not planted by man. There are very few primitive or virgin forests in the British Isles, but the degree of man's interference varies very greatly from casual extraction of mature timber, firewood, or coppice, to complete destruction. The number of tree species native to the British Isles is small, the status of some has been disputed, and there are several which, though they were originally introductions, now rejuvenate themselves as readily as do the natives.

It may be of interest briefly to consider our principal trees, since their ecological importance is great as dominants or main associates of climatic climax communities. It is the tree layer which forms the uppermost overhead canopy and controls, largely by the amount of light it allows to penetrate, the lower layers of shrubs and herbs.

Two species of oak are generally considered native : the stalked oak (*Quercus robur*) (Pl. 20a, p. 95) and the sessile or durmast oak (*Quercus petraea*, often called *Q. sessiliflora*) (Pl. 20b, p. 95). The stalked or sessile character refers to the acorns, not to the leaves. Many of our oaks have undoubtedly been planted, particularly to furnish hardwood timber for shipbuilding, but we know from the pollen analysis of post-glacial peats that oakwood has been a prominent and often predominant community at lower altitudes in many parts of the British Isles from some 7,000 to 8,000 years ago. The stalked oak is the dominant species on moist clays and loams but also occurs on poorer and even acid soils, as in Wistman's Wood on Dartmoor, which is generally accepted as a natural wood. When grown in the open the stalked oak develops a widely spreading canopy with zigzag branching. The sessile oak generally avoids the heavier clays and loams but becomes abundant or dominant on lighter, sandy, and other siliceous soils. When the two species grow near together they hybridize freely.

Only one species of beech (*Fagus silvatica*) occurs in the British Isles, outside a few gardens. As a native tree here, it is probably confined to the south of Britain, but has been widely planted elsewhere and flourishes even as far north as Aberdeen and Caithness. The species will grow well on a variety of soils—especially on deep, well-drained, fertile loam with a good supply of mild humus. It is very frequent on chalk and oolitic limestones and on such rocks, especially if they be covered by some overlying loam, it competes successfully with other British trees, partly because it spreads its feeding roots in very shallow soil, partly because of the deep shade it casts.

Two native birches attaining tree size are the silver birch (*Betula pendula*, also known as *B. alba* and *B. verrucosa*) and the hairy birch (*B. pubescens*). The birches grow on a great variety of soils but generally avoid chalk and the purer limestones. The hairy birch is much the commoner in the north and west and the silver birch in the south. Both species are common on sandy heaths, where they frequently hybridize.

There is only one species of ash (*Fraxinus excelsior*) native to the British Isles but this is widely distributed, except in the north of Scotland, and is often abundant. It is very variable in both leaf and fruit characters. On chalk and other limestones and on calcareous soils generally it may form pure ashwoods or be co-dominant with one or other of the oaks. It requires a good supply of mineral salts and avoids markedly acid soils.

The Scots pine or Scotch fir (*Pinus silvestris*) is a very hardy tree. It is certainly native in the Scottish Highlands where it is represented by a distinct variety (var. *scotica*) and was extremely abundant throughout the British Isles in early post-glacial times. Many of the pine woods on the heaths in the south of England have originated from plantations or gardens and now rejuvenate naturally (Pl. X, p. 175). It is, however, very probable that some at least of the English pine-woods may be partly descended from native stock. Generally speaking the Scots pine

John Markham

HAWTHORN, *Crataegus oxyacanthoides* (Rosaceae). Botany Bay, Middlesex. September 1945

John Markham

BLUEBELL, *Scilla non-scripta* (Liliaceae): robust form, not quite typical
Baltonsborough, Somerset. April 1946

flourishes best on siliceous soils, which are poor in mineral salts and acid in reaction.

The common alder (*Alnus glutinosa*) is widely distributed throughout the British Isles either locally in the wetter parts of woods dominated by other trees or along river, stream, and lake sides. In the wetter post-glacial times alder was more abundant than it is now as shown by the pollen records in peat. The draining of marshy ground has greatly reduced its occurrence.

The wych elm (*Ulmus glabra*) is widely spread sporadically in oak and other woods. On limestones it may form local societies in ash or ash-oak woods. It is more abundant in the north and west than in the south. Both the taxonomy and the phytogeographical status of the other elms have been matters of controversy. It is probable that several species are involved and that these have hybridized freely. Some elms are more or less sterile and some propagate freely from suckers. In various English counties elms are extremely common as hedgerow trees. Some of the microspecies (if such they be) appear to have general eastern or western ranges, but in some counties (central Oxfordshire, for example) there is a very great mixture of possible species and hybrids.

Species of *Sorbus* are very beautiful trees. The so-called mountain-ash or rowan (*Sorbus aucuparia*) (Pl. 22, p. 99) is widely distributed on lighter soils. It is particularly abundant in northern districts. The whitebeam (*Sorbus aria*) (Pl. XII, p. 191) and wild service (*Sorbus torminalis*) are mainly found in the south, the former being a common constituent of chalk woodlands and brushwoods. In the west of England, in Wales, and in western Scotland there are a number of *Sorbi* whose status is still uncertain. Some may be microspecies but many are probably hybrids or, at least, of hybrid origin. Other rosaceous trees are the crab (*Malus pumila* or *Pyrus malus*), which is not uncommon in the south of England and ranges in size from a shrub to a small tree ; gean or wild cherry (*Prunus avium*), especially in oak and mixed woods on limestone soils, as on the Chilterns ; the true cherry (*Prunus cerasus*) which may be native in southern England ; and the bird cherry (*Prunus padus*) which occurs at higher altitudes in the west and north.

There are two limes (*Tilia cordata* and *T. platyphyllos*) which are probably native in woods of the west country. Elsewhere in Britain limes are mainly planted trees or descendants of such.

Poplars are mostly introduced trees in the British Isles. The aspen

(*Populus tremula*) and the black poplar (*P. nigra*) may be native in some areas. The tree willows (*Salix fragilis* and *S. alba*) are commonly planted along rivers, canals, and streams and are frequently pollarded. Both may be native in some wet and marshy soils.

The yew (*Taxus baccata*) (Pl. 1, p. 34) is widely spread in England, Wales, southern Scotland, and Ireland, mainly on limestones and particularly on Chalk. Its use in the making of bows caused it to be widely planted in the Middle Ages.

The hornbeam (*Carpinus betulus*) has a restricted distribution as a native tree in the British Isles, being limited as such to south-eastern England. It is particularly abundant in Kent, Sussex, Essex, Middlesex, and Hertfordshire. Most often it is coppiced.

The holly (*Ilex aquifolium*) is the most widespread and the commonest of British evergreen trees. It ranges throughout the British Isles, except in some parts of the extreme north, and is particularly abundant in the mild damp areas of the west.

The common maple (*Acer campestre*) is usually coppiced or cut as a hedgerow shrub, but when left to grow to full size may form a tall tree. It avoids acid soils but is abundant on Chalk and other limestones.

There remain a number of certainly introduced trees which have been so widely planted and which, locally at least, reproduce themselves so successfully by seed that they often form components of or even dominate some of our existing woodlands. The sycamore (*Acer pseudo-platanus*) likes deep, loamy soil ; sweet or Spanish chestnut (*Castanea sativa*) avoids calcareous soils but flourishes on some southern sandy heaths where it ripens a proportion of its nuts though these are rarely of full size ; the horse-chestnut (*Aesculus hippocastanum*), belonging to quite a different family from that of the sweet chestnut, is widely planted as a decorative tree and seeds abundantly in the south and midlands ; the Turkey oak (*Quercus cerris*) forms most often a small and rather bushy tree but produces acorns freely and is spreading locally (as on Oxshott Heath in Surrey, Bladon Heath in Oxfordshire, and parts of the Poldens in Somerset) so that it may well become quite an important member of the British flora.

In the study of woodland or forest special attention must be given to layering or stratification. The community is dominated by the tallest layer of trees which may be close together so that their crowns touch and even interlock or so spaced that the crowns do not meet. The effectiveness of the tall tree canopy in controlling the layers of vegetation beneath depends on the kind of tree or trees present and on their spacing. Beneath the topmost stratum there may be another of shorter trees, which are either younger trees of the same species or trees of a different species, and beneath this a layer of shrubs often called the undergrowth. The layer of herbs in a wood is best termed the field layer. Pressed close to the soil there may be a ground layer of bryophytes and lichens. The presence or absence, and when present the number and composition, of the layers beneath the highest tree canopy depends in large part directly on this last modifying the physical factors of the environment. The nature of the soil, of course, plays its part, but is in its turn considerably influenced by both the root action and the leaf-fall of the trees. It has also to be remembered that there is a subterranean as well as an aerial stratification of plant organs. Excellent examples of " complementary societies " are found in woodlands. Thus, there is the soft-grass, bracken, and bluebell society found in open oakwoods on shallower soils where these three herbs live together in very close proximity, but because their underground parts are at different levels, and their aerial parts mature at different seasons, they completely fill the niche without undue competition for food and light. On deeper soils bracken rhizomes may compete with bluebell bulbs.

Oakwoods. There is no doubt that deciduous oak forest is the natural climatic climax for a very large part of the British Isles. The pollen-content of peat bogs, fossilized and semi-fossilized remains in submerged forests, documentary evidence from Roman times onwards, and the scattered remains till to-day of more or less natural oakwoods all point to this conclusion. Many of our British oakwoods are now coppice-with-standards. That is to say, there is a layer of shrubs (most often hazel) which is cut at periods of some 7 to 12 years and the shrubs shoot again from the cut stocks, while the tall trees are so spaced that their crowns do not or scarcely meet. Such woods are not now economic in their coppice and timber returns but have been mainly kept as cover for game birds. The stalked oak is frequently found on deeper soils in the midlands and south of England forming so-called " damp oakwoods." It is often very definitely the dominant tree, though the following are sometimes associated with it in one or another modification of oakwood : ash, birches, wych elm, hornbeam, aspen, maple, gean and alder. Beneath the tallest layer a second may be present with : holly, mountain-ash, service tree, crab, and sometimes yew. Many of the subsidiary trees are frequently coppiced. The shrub-layer is often dense, and while hazel (*Corylus avellana*) is the commonest oakwood shrub the following may also occur :

hawthorn (*Crataegus monogyna*)
blackthorn (*Prunus spinosa*)
guelder-rose (*Viburnum opulus*)
privet (*Ligustrum vulgare*)
spurge laurel (*Daphne laureola*) (Pl. 27, p. 114)
dogwood (*Cornus sanguinea*)
elder (*Sambucus nigra*)
spindle-tree (*Euonymus europaeus*)
willows (*Salix capraea*, *S. aurita*, and *S. atrocinerea*)

Ivy (*Hedera helix*) and honeysuckle (*Lonicera periclymenum*) (Pl. 25) are common climbers in stalked oakwood. Brambles (*Rubus* spp.) are frequent as clumps, clamberers, or low trailers. Toothwort (*Lathraea squamaria*) (Pl. 21, p. 98) often occurs parasitic on hazel roots. In coppiced oakwoods, especially two or three years after coppicing, extensive carpets of herbs, often with one species dominant over considerable stretches, are a striking feature. Everyone is familiar with these in their prevernal (March–April) or vernal (April–May) aspects, as they flower before the oaks develop their leafy canopy and cut off light. Then appear the sheets of colour provided by the flowers of the bluebell (*Scilla non-scripta*) (Pl. 24, p. 107), wood anemone (*Anemone nemorosa*) (Pl. 26, p. 111), primrose (*Primula vulgaris*), violets (*Viola* spp.), and lesser celandine (*Ranunculus ficaria*). Dog's mercury (*Mercurialis perennis*) has inconspicuous greenish flowers (male and female on different plants) but often forms pure societies of considerable extent. Other herbs found in stalked-oak woods are wood sanicle (*Sanicula europaea*), ground ivy (*Glechoma hederacea*), enchanter's nightshade (*Circaea lutetiana*), yellow archangel (*Galeobdolon luteum*), wild strawberry (*Fragaria vesca*), wood spurge (*Euphorbia amygdaloides*), rosebay willow-herb (*Epilobium angustifolium*), and, particularly in the west, foxglove (*Digitalis purpurea*). Where there is loose soil or decaying organic debris the stinging nettle (*Urtica dioica*) forms local societies often to the exclusion of all other herbs. Mosses and fungi are generally abundant, the latter being most conspicuous in the autumn. Bugle (*Ajuga reptans*) is often conspicuous in damp places with heavy soil. The soft grass (*Holcus mollis*) is one of the commonest oakwood grasses, forming carpets of grey-green foliage. Bracken (*Pteridium aquilinum*) may form

John Markham

HONEYSUCKLE, *Lonicera periclymenum* (Caprifoliaceae). Pennard, Somerset. October 1946

PLATE 26

Cyril Newberry

very extensive societies to the exclusion of other species. Throughout the summer months its green fronds (all above ground is leaf since the stems are underground as slender, long, branched rhizomes) so shade the soil that lower-growing plants cannot compete with it. There is also some evidence that the dead decaying fronds of autumn and winter make the ground unsuitable for the establishment of other species.

Sessile-oak woods are especially characteristic of the older siliceous rocks of the west and north of England, of Wales, and of Ireland but the sessile oak also occurs on soils with relatively high acidity in the midlands and south-east of England. Thus sessile oak–hornbeam woods of Hertfordshire have been described in detail. In these woods hornbeam is a coppiced shrub while hawthorns (*Crataegus monogyna* and *C. oxyacanthoides*) (Pl. 23, p. 106), brambles, and willows are frequent, and guelder-rose, alder buckthorn (*Frangula alnus*), service tree, and holly though rarer are characteristic. The field layer is fairly rich in species but rather scanty in individuals. Bramble and bracken societies are well developed. Other good examples of sessile-oak woods are to be found in the Pennines, in the Malvern Hills, in the Forest of Dean, in Wales, in Ireland (as at Killarney), and in the valleys and glens of the Scottish Highlands. The associated shrubs of the undergrowth and the herbs of the field layer vary locally but the general list is much the same as for the stalked-oak woods. The abundance of bryophytes and lichens is particularly marked in the damper climate of the western oakwoods.

On sands and gravels, of different geological ages, in the south-east of England, mixed oakwoods are well developed in general association with a heath flora. Both the stalked and sessile oaks occur and hybrids between them are frequent. The name "oak-birch heath" is sometimes applied to variants of this community in which birches also grow. Beech and Scots pine are frequently found with the oaks and birches. In between the trees, and sometimes beneath them, different forms of heath vegetation cover the ground—heath proper with heather (*Calluna vulgaris*) and common heath or bell heather (*Erica cinerea*); stretches of bracken; grass heath with bent grasses (*Agrostis* spp.), sheep's fescue (*Festuca ovina*), and wavy hair-grass (*Deschampsia flexuosa*); and gorse or mixed thorny scrub. The developmental status of these mixed oakwoods and oak-birch heaths may not always be the same and it is rather surprising that more detailed studies have not been made on them. Many of them would probably develop into tall woodland were the biotic factors of cutting, burning, and rabbit attack to cease action.

Beechwoods. In contrast to the oaks and birches, there is only one species of beech in the British flora, the common beech (*Fagus silvatica*). This forms extensive woodlands on the Chalk of the North and South Downs and the Chiltern Hills and on the Inferior Oolite (Jurassic limestone) of the Cotswolds. The largest areas are in the Chiltern Hills of Buckinghamshire and southern Oxfordshire and their preservation is doubtless connected with the chair-making industry which is still carried on in the woods, the craftsmen being termed " bodgers." Beech is intolerant of water-logged soils but competes successfully with oak on the shallow chalk soils (rendzinas). Beech is often characteristic of chalk slopes forming woods known as " hangers." The tree layer is almost entirely pure beech with a horizontally spreading root system from which anchoring roots enter fissures in the chalk below. Occasionally ash trees reach the canopy and on the Chilterns gean (*Prunus avium*) is associated with the beech. Yew (*Taxus baccata*) (Pl. 1, p. 34) and holly (*Ilex aquifolium*) sometimes form a second tier. The beech is a very leafy tree and the density of the foliage with the tendency to a horizontal spreading of branches and leaf mosaic means that very little light penetrates the canopy. As a consequence, the undergrowth is very sparse or absent and the field layer either consists of a few highly specialized plants or is quite absent. Other factors inimical to the development of a herbaceous carpet are the deep leaf litter and leaf mould and the great development of superficial beech roots with their strikingly numerous mycorrhiza (shortened roots associated with fungal hyphae) which can be readily seen by removing a few inches of superimposed dead leaves. In dense beechwoods saprophytes (plants living on dead organic matter) and mycorrhizal plants may be the only herbs. Thus of saprophytes there are the yellow bird's-nest (*Monotropa hypopitys*) and the bird's nest orchid (*Neottia nidus-avis*), while mycorrhizal plants are represented by two helleborines (*Cephalanthera latifolia* and *C. longifolia*) and tway-blade (*Listera ovata*). Where the woods are more open, or in clearings in such woods, a richer flora and a denser field-layer may occur, with such plants as wood sanicle (*Sanicula europaea*), sweet woodruff (*Asperula odorata*), wood violets (*Viola* spp.), archangel (*Galeobdolon luteum*), wild strawberry (*Fragaria vesca*), dog's mercury (*Mercurialis perennis*), enchanter's nightshade (*Circaea lutetiana*), and a number of woodland grasses, especially melic grass (*Melica uniflora*) and wood meadow-grass (*Poa nemoralis*). The coralroot (*Dentaria bulbifera*) is abundant in some Chiltern beechwoods where it reproduces by axillary bulbils. The ground layer of mosses, liverworts and Fungi (toadstools) is characteristic. The moss *Hypnum cupressiforme* is often abundant on the bases of the beech trunks and *H. molluscum* on the soil and on stones (particularly in the Cotswold beechwoods). The remarkable moss *Leucobryum glaucum* occurs in tufts which may become detached from the substratum. The liverwort *Metzgeria furcata* occurs on the trunks. Common species of larger Fungi are *Collybia radicata*, *Mycena pura*, and *Hygrophilus eburneus*.

Much of our chalk and Jurassic limestone country is now arable or grassland. Two seres (sequences of developmental stages) have been distinguished leading to beechwoods when certain biotic factors (including rabbit grazing and pasturing) are prevented from acting. These both pass from grassland through scrub stages to tall beechwood but, in one, juniper (*Juniperus communis*) is the dominant shrub of the intermediate stage which beech invades directly, while in the other, hawthorn is the main component of a brushwood which is succeeded by ashwood before beech enters.

Beechwoods occur, in the south of England, not only directly on the Chalk and Jurassic limestone but also on loamy soils, especially on the chalk plateaux covered with Tertiary deposits and "clay-with-flints." In these woods, stalked oak, gean, sycamore, and whitebeam (*Sorbus aria*) may be associated with the beech in greater or less abundance. On the South Downs, yew and holly may form a second layer but there is not usually a second tier in the plateau beechwoods of the Chilterns. The bramble (*Rubus fruticosus* agg.) is very common as a low sprawling shrub. Where bramble is absent, wood-sorrel, bluebell, dog's mercury, and other herbs may occur. There are well-known beechwoods in the vicinity of London on gravelly soils such as those in parts of Burnham Beeches and Epping Forest.

Yew woods (Pl. 28, p. 115). The yew, as we have seen above, is frequent in some types of beechwoods. Much more rarely it forms pure woods. When it does, as on parts of the North and South Downs (there is a small but interesting example on a part of Box Hill, Surrey), it casts such a dense shade all the year round that it suppresses nearly all other plants.

Ashwoods. Ash is frequently a dominant in a successional stage leading to beechwoods but on a great many of our limestone outcrops, outside the native range of the beech, ashwoods form the climax vegetation on a variety of calcareous soils. The ash casts a much lighter shade than even the oak and in ashwoods there is usually developed a rich undergrowth of shrubs, or a field-layer of herbs, or both. Trees which accompany the ash include wych elm (*Ulmus glabra*), aspen (*Populus tremula*), yew (*Taxus baccata*) (Pl. 1, p. 34), whitebeam (*Sorbus aria*) (Pl. XII, p. 191), and maple (*Acer campestre*). In the shrub layer are often to be found : dogwood (*Cornus sanguinea*), spindle-tree (*Euonymus europaeus*), buckthorn (*Rhamnus cathartica*), privet (*Ligustrum vulgare*), willows (*Salix capraea* and *S. atrocinerea*), roses, and brambles. Ivy (*Hedera helix*) is often abundant. Honeysuckle (*Lonicera periclymenum*) (Pl. 25, p. 110) and old man's beard (*Clematis vitalba*) are locally common. The field-layer has a very rich flora which varies both with the thickness of the undergrowth and with the dampness of the soil. On damp soil, societies of ramsons (*Allium ursinum*) (Pl. XI, p. 190) and of lesser celandine (*Ranunculus ficaria*) are found, and other interesting species include woodruff (*Asperula odorata*), globe-flower (*Trollius europaeus*), and valerians (*Valeriana officinalis* and *V. dioica*). On drier soils there are communities of dog's mercury, often with moschatel (*Adoxa moschatellina*) and ground ivy (*Glechoma hederacea*). Mosses and liverworts are often abundant in ashwoods.

Pinewoods. It is usual to consider British pinewoods as falling into two very distinct groups : primitive woods of Scots pine (*Pinus silvestris* var. *scotica*) in Scotland and naturalized woods of *Pinus silvestris* var. *typica* (Pl. X, p. 175) on sands in the south of England. It is probable that at least one other group will have to be added when pine woods in the north, and perhaps other parts, of England have been studied in detail. The Scottish pinewoods are now scattered remnants of formerly extensive forests. The density of the tree growth in them varies greatly. When this is high the undergrowth and field-layer are sparse or absent, but when the trees are well-spaced, heather and bilberry may make a close cover. Juniper (*Juniperus communis*), cowberry (*Vaccinium vitis-idaea*) (Pl. 19, p. 94), crowberry (*Empetrum nigrum*), and some grasses are also sometimes present. Of rare species occurring in Scottish pine-forests mention should be made of lesser twayblade (*Listera cordata*), *Goodyera repens* (another orchid), winter-greens (*Pyrola* spp. and *Moneses uniflora*) (Pl. XIII, p. 194), coral-root orchid (*Corallorrhiza trifida*), *Linnaea borealis* (Pl. XIV, p. 195), and chickweed winter-green (*Trientalis europaea*) (Pl. 18, p. 91). Mosses

F. Ballard

SPURGE LAUREL, *Daphne laureola* (Thymelaeaceae)
Wootton Wood, near Woodstock, Oxfordshire. April 1945

Brian Perkins

Chalkland vegetation, Box Hill, Surrey. View overlooking the hair-pin road, with *Crataegus* and *Taxus*. October 1946

are often extremely abundant and in some pinewoods bog-mosses (*Sphagnum* spp.) play a very prominent part in the ground flora.

The pinewoods of the south of England are familiar to many Londoners as forming the dominant tree vegetation on well-known areas of sandy or gravelly soil in Surrey (Oxshott, Esher, Wisley, Leith Hill, etc.) (Pl. 29, p. 122). The pine in such habitats rejuvenates naturally and stages in succession leading to tall pinewood can be studied in many places. Where the pines grow thickly, little light penetrates the evergreen canopy and the ground beneath the trees is covered with a deep layer of fallen " needles " (dwarf branches with two leaves) which slowly decay to form a very acid mould. In the autumn a rich Fungus flora may be found in these pinewoods but otherwise the ground is bare except for a few mosses of which the tufted moss *Leucobryum glaucum* (also found in some beechwoods) is one of the most interesting. Where the trees are spaced, heath vegetation penetrates with heather, grasses, such as *Deschampsia caespitosa* and *Molinia caerulea*, and even gorse.

Birchwoods. These are common on sandy and gravelly heaths in southern England. Frequently they represent stages in a secondary succession (often after heath fires), leading to oak-birch heath or to oakwoods. Sometimes they are relicts of oak-birch woods from which the oaks have been extracted. The birches on the heaths and commons round London are a mixture of our two tree species with hybrids between them. Casting a very light shade the birch allows the development of an abundant undergrowth and ground flora and the physical and chemical nature of the soil, which has usually an acid reaction, limits these to heath plants. Birches, particularly the hairy birch, are important woodland dominants in the north of England and in Scotland. Birchwoods here form the altitudinal forest limit in the mountains in Perthshire and Inverness-shire at about 600 m. (2,000 ft.). Grasses are often dominant in the ground layer of these northern birchwoods.

Alderwoods. These have without doubt been considerably reduced in extent within historic times as a result of drainage of marshlands. The alder is widely distributed throughout the British Isles and flourishes exceptionally well in the wetter climate of the west. However, some of the most extensive alderwoods now existing are to be found in East Anglia where the alder dominates many of the fen woods or carrs. It grows elsewhere in water-logged soil, when this is not too acid, in valleys and by the sides of rivers, streams, and lakes often in association with other trees such as ash, oak, and birch. Associated shrubs commonly found are :

willows
buckthorn (*Rhamnus cathartica*)
alder-buckthorn (*Frangula alnus*)
black currant (*Ribes nigrum*)
gooseberry (*R. grossularia*)
privet (*Ligustrum vulgare*)
guelder-rose (*Viburnum opulus*)

The herb-layer is largely composed of marsh plants such as :

yellow flag (*Iris pseudacorus*)
sedges (*Carex* spp.)
marsh marigold (*Caltha palustris*)
meadow-sweet (*Filipendula ulmaria*)
fern (*Dryopteris thelypteris*)

The stinging-nettle (*Urtica dioica*) is sometimes abundant and there are numerous mosses and liverworts.

Mountain Communities

The natural altitudinal limit of forest probably lies between 1,500 and 2,000 ft. for most parts of the British Isles except the extreme north. Above this limit one finds only low shrubs and herbs, many of them belonging to arctic-alpine species. Much of the surface is broken, rocky, and stony and there are many steep slopes with more or less unstable scree. In some areas scree becomes gradually covered with vegetation of a heath type, through a series of stages involving accumulation of humus and stabilization of the surface. The final stage may be sessile-oak wood when the scree is below the forest limit. Above this limit grasslands may be formed, often dominated by sheep's fescue (*Festuca ovina*). On mountain top plateaux, crests, peaks, and ridges the vegetation is usually low, sparse, and open. In all of our mountain areas lichens and bryophytes play a very important part in the successional stages (which are necessarily few) and in the climax communities. Thus there is one community, known as "*Rhacomitrium*-heath," in which the substratum is more or less covered by woolly fringe moss (*Rhacomitrium lanuginosum*).

Special mention must be made of one interesting fact. The richest alpine-arctic flora, and the most luxuriant vegetation, usually occurs, in the Scottish Highlands, along the limited exposures of calcareous schists. This, for example, is well seen on Ben Lui and on Ben Lawers.

Artificial Communities

Space will permit of only a brief reference to cultivated crops, though their ecological and economic importance is often great. Many of the communities already described have been more or less modified by human activities and some are maintained in their existing condition as an immediate result of the human biotic factor. Besides these partly natural and partly artificial communities, a very large proportion of the surface of the British Isles, particularly in England, is covered with farms, gardens, and plantations, where cultivated crops are the dominant plants. The naturalist should not attempt to ignore this fact because he can learn so much from the practices of agriculture, horticulture, and forestry. Only a very few of the interesting problems associated with crops can be mentioned here and further details should be sought in other volumes of this series.

Arable farmland. As with natural vegetation, so with cultivated crops, it is the climatic factors that act as major controls. In the British Isles, the cereals grown are mainly wheat, barley, and oats, corresponding with the temperate nature of the climate. Oats can stand colder and wetter climates than wheat and barley and therefore are the commonest, or only, cereals in parts of the north and west. "Roots" include swedes, turnips, mangolds (mangel-wurzels), sugar-beet, carrots, and potatoes. The last (the crop is really one of stem-tubers) is grown all over the British Isles but the best seed tubers come from the north where the climatic conditions do not allow the vectors (carriers or distributors) of potato virus diseases to flourish. Temporary leys of grasses, clovers, sainfoin, lucerne, vetches, kidney-vetch, etc., whether pure or in mixtures, are increasingly grown and their employment is likely to be correlated with considerable changes in farm practice. There is often something particularly attractive to the botanist in the cultivation of special crops such as hops (in Kent, Sussex, and Herefordshire), flax (in Northern Ireland and increasingly in some parts of England), bulbs (in Lincolnshire), and tobacco (in Hampshire).

While climate largely determines in a broad sense the crops which can or cannot be profitably grown, edaphic (soil) and biotic factors often determine which is grown in the greatest quantity and even more definitely determine the varieties. Deep clays, porous sands, shallow chalk soils, and so on cannot all be expected to grow the same crops any more than one would expect them to have the same flora or the same plant communities. Farming methods, including tradition, proximity to markets, and availability and price of seed, are biotic factors, in the broad sense, that influence the exact kind of crop grown.

Crop rotation opens up a whole subject full of ecological implications as does also the study of diseases and pests of cultivated crops. The weed flora has a double interest, apart from its practical importance. A large number (very probably a majority) of our weeds of arable land are not strictly native species but have been introduced with the seeds of crop plants. A good example is the weed speedwell (*Veronica persica*). Much remains to be done in tracing the history of our common weeds but many seem to be of southern (especially Mediterranean) origin. The second interest of weeds concerns their adaptation to the peculiarities of an environment that is constantly being disturbed and in which there is severe competition with a sown and temporarily dominant crop. Some examples of such adaptation are given in a later chapter, and here it will suffice to mention that, in general, weeds of arable land show many of the characteristics found in the colonizers of an early stage in plant succession.

Artificial grasslands. Grasslands may be temporary or (more or less) permanent. Temporary leys are best considered as part of the rotation on arable land. Permanent grassland may be used for grazing (pastures) or for hay (meadows) or alternatively for both. In permanent pastures a majority of the species are probably native and the floristic composition depends at least as much on treatment and on the soil as on climate. Floristic analysis, counting and recording the numbers of every species in from 10 to 50 sample plots of 1 m. square, will enable instructive comparisons to be drawn between one field and another and differences can then be correlated with environmental differences. Grasslands, of course, contain many plants that are not grasses and the relationship between these " forbs " (herbs other than grasses) and the dominant grasses is important. Some forbs are weeds, in the economic sense of lowering the quantity or quality of forage or hay, but others, such as ribwort plantain (*Plantago lanceolata*), are valuable in supplying nutrients (including minerals and vitamins) in which the grasses are deficient.

Gardens and orchards. Much of what has been said for arable farming applies, with differences of scale and crops, to horticulture. Market gardening is in some ways intermediate between farming and the cultivation of a private garden. Three differences between farming and gardening are obvious. Firstly, many more kinds of plants can be and often are grown in a garden than on a farm, but the crops are smaller. Secondly, in many gardens an artificial microclimate is provided by hothouses, cool greenhouses, or frames and, on a small scale, soils may be made or imported. Thirdly, a garden is often not cultivated solely or even chiefly for its economic return in financial terms. On the other hand, such matters as rotation, diseases and pests, and weeds are not fundamentally dissimilar. To the naturalist a garden is of particular importance in that it enables him to experiment with plants and to keep them constantly under observation.

Plantations. Planted woodlands are, to the ecologist, experiments with artificial climax communities or constituents of such. Their success or failure and their effect on associated plant and animal life offer for investigation problems whose solution may have significance in the study of natural woodlands. We have already noted that a number of trees, such as the sycamore, Turkey oak, and sweet chestnut, reproduce and locally spread almost as if they were native trees. They are not natives of our flora, either because they have not reached the British Isles by natural extension of range or because they could not compete with native trees in full natural competition. On the other hand, the plantations of spruces, larches, Douglas fir, and various pines do not extend their boundaries and remain only as cultivated crops. Their reactions to our climatic and soil conditions, their diseases and pests, and the small number of plants often associated with them (the number of animals is said to be greater) are matters of biological interest.

Ruderal communities. The widespread and varied activities of man have led to the establishment of many " waste places," which though most often of small individual extent are extremely varied. Since they are usually closely associated with human habitations they become occupied not only with relicts of the native flora but also with numerous weeds and casual introductions. Waste places near houses often have communities of wall barley (*Hordeum murinum*), goutweed (*Aegopodium podagraria*), or stinging nettle (*Urtica dioica*). The remarkable spread of the Oxford ragwort (*Senecio squalidus*, a very polymorphic species, of Sicilian origin) has not yet been fully traced or explained, and its range is still extending. It is supposed to have spread from Oxford first along the Great Western Railway, but its spread has been very unequal in different directions. The author failed to establish it artificially by either transplants or fruits on walls at Woodstock, 8 miles north of Oxford, forty years ago, and has only recently found one plant there, and that in the G.W.R. station yard. It was established at Oxford as long ago as 1794 but its phenomenal spread to the London area and to many other parts of the south and midlands has occurred in the last half century. With another alien ragwort (*Senecio viscosus*) and the rose-bay willow herb or fireweed (*Epilobium angustifolium*) it has become a common plant on bombed sites in London. This and a great many other ruderals (plants growing in waste places) deserve careful investigation as to their ecology and genetics. The spread of the slender speedwell (*Veronica filiformis*) (Pl. 30, p. 123) should be recorded in order to contrast its history with that of the now common weed speedwell (*Veronica persica*).

Dockyards, railway-sidings, and rubbish tips may not appear very inviting botanical areas but, again, their flora may be regarded as the result of unintentional experiments in introduction. Most of the very large number of species introduced through commerce (with grain, fruits, wool, etc.) do not survive, but the causes of their non-survival are not always easy to determine and may range from failure to set viable seed to failure to compete with other species or with the climate.

CHAPTER 10

VARIATION

VARIETY, it is sometimes claimed, is the spice of life. Plant life certainly cannot be accurately described as uniform or monotonous. In biology, however, the word "variation" is most often used to indicate difference from some standard—from some normal or mean form or behaviour—or the differences shown between individuals (or organs) grouped in a single class. If we limit our discussion to these aspects there is an enormous field to cover and almost innumerable instances from which to select examples. Before considering variation in our British flora there are several general concepts which it will be advisable to note.

Firstly, variation may be continuous or discontinuous. By continuous variation is meant that a series can be found in an organ or character showing every gradation from one extreme of development to the other with no marked break anywhere in the series. In discontinuous variations, on the other hand, there are marked breaks, an obvious and easily describable or measurable difference separating one variant from another. Actually there is every intermediate between what is markedly discontinuous and what appears to be continuous variation and it does not necessarily follow that the continuous series we arrange is the sequence in which the variants arose in the course of evolution.

Secondly, variation may be quantitative or qualitative. That is to say we can record the characters and determine the variation quantitatively by measurements of height, length, width, volume, or weight, or, at least with our present knowledge and technique, we can only or best describe it in words. Again there is no absolute distinction between quantitative and qualitative characters, the difference between them being more than anything else a question of the scientifically most convenient and practical method of obtaining and analysing the facts.

Thirdly, we have to distinguish between variations due to different environments acting on the same inherent constitution and those due to different constitutions maturing under the same or different environments. It is necessary here to interpolate what may seem a rather verbose explanation, but, since it concerns a matter of the utmost biological importance on which there is much misunderstanding, an attempt must be made to make clear what is meant by " fluctuations " or " environmental modifications " on the one hand and " genetic variations " on the other. To do this we have to introduce the concept of " inheritance," which in biology means that which the offspring receives from parents, and the concept of " ontogeny," or the developmental life-history of an organism. The vast majority of plants and animals are sexually reproduced and one can reasonably say that a new individual is formed at and as a result of fertilization. Fertilization is the union of a male cell with a female cell. These sexual cells, or gametes as they are termed, have contents and potentialities different from those of the body cells, but details of these peculiarities need not detain us at present. The immediately essential point is that fertilization is the fusion of two microscopic portions of living matter, very small particles of jelly-like protoplasm, and that the new individual starts its development as a similar naked tiny mass of protoplasm with none of the visible organs or characteristics of the adult organism. All of these appear in due course, in regular sequence, with or without intervals of time, during the development (ontogeny) of the fertilized female gamete (zygote) into the adult organism. In other words organs and characters are not inherited : they develop afresh in every individual. Yet we know that figs do not grow on thistles even if fig trees and thistles be growing in the same garden. There is control by something which, within suitable environments, determines the development of this or that organ, structure, quality, or behaviour. These " somethings " are termed factors or genes and there are good reasons for believing that they are actual material particles. The genic concept of inheritance will perhaps be made clearer by taking a very human example. It is often said of a new-born babe that " he has his father's nose." Obviously the phrase taken literally is sheer nonsense. The father does not take off his nose and fix it on his son as part of the reproduction process. The father or son has each his own nasal organ. The underlying meaning is that certain characters (usually of shape) are similar in the corresponding organ of father and son and it is correctly assumed that these similarities are due (within limits) to the factors contributed by the paternal sex cell at fertilization. A similar explanation can be applied to such phrases as a plant " inherits " blue flowers, or hairy leaves, and so on.

We have to note, then, that without the controlling or determining

Brian Perkins

Vegetation on sandy soil, Leith Hill, Surrey. South-west slope, with *Pinus silvestris*, *Pteridium aquilinum* and *Ilex aquifolium*
October 1946

John Markham

SPEEDWELL, *Veronica filiformis* (Scrophulariaceae). Glastonbury, Somerset. May 1946

genes such and such a character will not develop, but even if the genes be present the character will not appear unless there be also a suitable environment or environmental complex for its development. A simple example will make this clear. A gorse plant has the factors for producing spiny branches and reduced scale leaves under the conditions normal to our British heaths, but a gorse plant grown under a glass bell-jar with plenty of water and in a saturated atmosphere will produce only elongated non-spiny branches and well-developed leaves with three leaflets.

While it is often necessary to distinguish variation due essentially to differences in genic or factorial make-up from that due immediately to environmental differences to which the individual has been exposed in the course of its development, it is equally important to remember that the inherited factors and the environment (internal and external) are always interacting and that any character, organ, or individual is the summated result of this interaction. The genic constitution is normally fixed at fertilization, while the environment is constantly changing. The genic constitution is inherited, the external environment is not, in a strictly biological sense. Only experiment can prove, though observation can often suggest, whether a variation is basically genic or environmental. Further, there is often an overlapping, even in controlled experimental research and still more frequently in nature, of character-expression due to different inherited factors and to different environments respectively. Suppose plants, as, for example, the knapweeds (*Centaurea nigra* varieties) (Pls. 44, p. 211, and 45, p. 218), are being studied for height of stems, and that A and B have genes for tall stems, C and D genes for dwarf stems. If A and C be grown under good conditions of soil, moisture, temperature, and so on, and B and D be starved of food and water and kept at lower temperatures, the order of stem height, tallest first, may well be A, C, B, D. Nevertheless, simple experimentation with well-known and obviously reasonable precautions will nearly always enable genic variation to be distinguished from environmental.

Variation is such a widespread phenomenon in plants, affecting all organs and all functions, that in this and the following chapters only a relatively few selected examples can be given. Apart from selecting as wide a range as possible to illustrate different kinds of variation, in different groups of organisms, and involving different organs and functions, it is of secondary importance how we arrange our examples.

There is some advantage in a systematic sequence, but this would involve repetition or too much reference to earlier or later paragraphs, so that a classification based firstly on kind of variation and secondly on the organs and groups involved has been attempted for the main part of this chapter.

Fluctuations or Environmental Modifications

The importance and interest of plasticity, or the range of development in a character with varying environmental conditions, is not always realized. Firstly, it not infrequently causes difficulties in determination and the naturalist must learn to recognize the common "habitat forms" of species, i.e. modifications constantly associated with special environments. Recent research has shown clearly that many so-called varieties, and even some described subspecies and species, are no more than environmental modifications. Secondly, there is the possible survival-value of being able to live and reproduce under a more or less wide range of environmental conditions, a subject discussed more fully in the next chapter. Though fluctuations of a morphological nature are the easiest to illustrate, it must be remembered that they are correlated with fluctuating physiological behaviour and that this last may also occur without immediate visible changes of structure. That some species are restricted to a single community while others are components of several or even many different communities may be partly due to differences in plasticity. In the following account of fluctuations, headings are given under which the reader can add many more examples from his own investigations to the few chosen as illustrations.

Fluctuations in different parts of the same individual. In ivy, lobed leaves occur on purely vegetative branches especially when these are climbing up a support by their adventitious roots, while entire leaves occur on projecting branches which flower and fruit. In the holly, the leaves on the lower branches have many more marginal prickles than the leaves on branches high up in the tree. Herbs, too, frequently show differences in shape and size between lower and upper leaves. The harebell (*Campanula rotundifolia*) has very narrow grass-like leaves on its flowering stems and seeing only these one may well wonder why the Latin epithet *rotundifolia* was given. The lower leaves, however, have roundish blades on distinct stalks, though the change from these to the narrow kind is not always abrupt. The snowball shrub (*Symphoricarpus albus*) shows a great range in leaf-shape.

Another interesting fluctuation of corresponding organs in different parts of the same individual is that of sun and shade leaves. Thus, beech trees often grow in such a situation that some branches are always shaded while others are fully exposed to the sun and the winds. The leaves on the former relative to those on the latter are generally larger, more flaccid, thinner, and of a paler green colour. Comparisons of transverse sections of sun and shade leaves may be made under a microscope or even a high-powered hand-lens.

Reference should be made here to "juvenile forms." The first leaves and branches above the cotyledons in the seedlings of many plants are different in form and size from those borne on mature branches. Gorse seedlings illustrate this well. The most striking examples of juvenile forms are, perhaps, those of certain cultivated plants, including some conifers. Gardeners have propagated seedling shoots by cuttings and these have sometimes grown into fairly large plants which retain the juvenile characters indefinitely. Such permanently juvenile forms of some junipers have been given the name Retinospora and do not flower or fruit.

Pollarding, cutting, and suckering. Everyone is familiar with pollarded willows, whose shape is changed by the early removal of the top part of the main trunk and the consequent development of a crown of branches of more or less equal size, these branches being again cut after an interval of about 10 years. Good examples of old pollarded beeches are to be seen at Burnham Beeches. Hazel and sweet chestnut are frequently coppiced. Hedgerows are most often drastically cut back periodically. Such management does not kill the tree or shrub but it frequently induces changes in its organs. Pollarding or lopping of branches and also damage to the root system stimulates the production of suckers in poplars and elms. In the former the leaves are often very much larger on sucker shoots than on normal branches, while in some elms they are smaller. In the ash, cutting back of branches sometimes produces fasciations and strange leaf-forms to appear from buds developing subsequently and the black mulberry can show the same phenomenon. Probably the cause of these abnormalities is a diversion of or an upset in the normal functioning of growth-controlling substances or growth-hormones, rather than a direct disturbance in food-supplies.

General size and general habit. Seeds of plants often reach, and germinate in, a variety of habitats. " Stony ground " and " good ground " may be juxtaposed and a crowd of seedlings jostle one another over both. Then great differences in size and in development of all parts can be observed—minute poppies, tiny clovers, depauperated plantains strive to survive and to flower on hard dry terrain while plants of the same species are tall and lush on neighbouring rich soil. A series of plants of common brome grass (*Bromus hordeaceus*) was collected in Surrey within a few square yards with spikelets ranging from one to fifty per plant, with every intermediate figure represented at least once. The degree of spine development in sloe or blackthorn (*Prunus spinosa*) and of prickle development in thistles varies greatly with the habitat, being less in damp and shady and greater in drier and more exposed places. Species normally aquatic sometimes occur on mud or on dry land and the plants then assume quite different habits of growth. Good examples can be observed in mare's-tail (*Hippuris vulgaris*), water-violet (*Hottonia palustris*), starworts (*Callitriche* spp.), water buttercups (*Ranunculus* spp.), and water bistort (*Polygonum amphibium*).

Developmental rhythm. It is well known that there are considerable differences in the normal developmental rhythm from species to species and even from variety to variety of the same species. Many of these differences are " fixed " genetically within limits. Nevertheless, there are narrower or wider fluctuations which can usually be correlated with environmental conditions. The name phenology is given to the study of recurring annual and seasonal variations such as those in shooting of leaf-buds, opening of flowers, and fall of leaves. The exact dates of occurrence of these and of some other seasonal phenomena are controlled in nature largely by temperature and rainfall and vary from year to year. The number of flowers, fruits, and seeds produced also depends, within the limits of genetic control (heredity), on the environment. It must be remembered that for some perennial plants, especially some trees and shrubs, the weather of the previous year may be an important factor in determining the amount of blossom produced. Good " mast " or fruiting years occur in the beech, oak, ash, maple, holly, and other trees and shrubs at irregular to subregular periods. There is as yet no fully satisfactory casual explanation of these variations, but both internal rhythmic factors and the environment appear to play a part.

The individual plant shows a normal rhythm of vegetative growth

F. Ballard

FLEABANE, *Pulicaria dysenterica* (Compositae). Kennington, Berkshire. September 1944

Brian Perkins

FLEABANE variety, *Pulicaria dysenterica* var. *hubbardii* (Compositae)
Islip, Oxfordshire. August 1946

followed by a reproductive phase. The normal rhythm is in part controlled by the environment and fluctuates with this. Light, temperature, water, and food supplies are variously involved but there is sometimes a rather clear-cut action of the last. Thus, in many species, heavy feeding with nitrogenous materials will encourage vegetative growth and delay flowering. It has been shown that the ratio between carbohydrates formed by the green parts and nitrogenous foods from the soil is often strongly correlated with flower production, a high ratio resulting in increased flowering and a low ratio in reduced flowering but increased vegetative development.

Teratological phenomena. The study of extreme divergences from the normal, that is of obvious abnormalities, is called teratology. Some abnormalities are environmental modifications, others are correlated with genetic changes. Fasciations (Pl. XV, p. 202) are the growth in union of what are normally a number of separate organs of the same kind. The commonest is a fusing and the common growth as one of a number of branch primordia (growing points). This usually produces a queer flattened branch obviously of composite structure. It sometimes results from pollarding, insect damage, disease, or even over-feeding. Other striking abnormalities are gall-formations due to insects, as in the numerous galls of oaks, rose bedeguars (or bedegars) which are sometimes called robin's pincushions, and so on. Fungus diseases produce abnormalities in grasses and other plants. If a large population of white clover, especially one on poor and rather dry soil, be examined, some plants will probably be found producing green flower-heads and in these abnormalities affecting all parts of the flower will be found. The production of these abnormal heads is said to be due to mites. Curious compact inflorescence galls are commonly produced on rushes by the insect *Livia juncorum* (Hemiptera-Homoptera, Psyllidae).

Special methods of study. As indicated in the above paragraphs, a great deal that is of interest can be learned about fluctuations by careful observation in the wild. Much more of value, however, can be discovered by methodical examination of whole populations or of large random samples of these. Such studies usually necessitate statistical methods. While these involve some concentration of effort in making and working out measurements or other scorings, the essential formulae are not difficult to understand and use. An outline account of the most useful statistics with which naturalists should be acquainted is given in Appendix C.

Of very great importance also is the experimental study of variations. For the moment, we are concerned only with fluctuations. Simple cultural experiments—such as growing different stocks under similar conditions—will often suffice to distinguish fluctuations from genetic variations. As an example of more elaborate, carefully controlled, and fully recorded experiments a brief outline is given below of the British Ecological Society's Transplant Experiments, at Potterne, Wiltshire, which extended over 13 years (Pl. XVI, p. 203).

These experiments were designed to test the interactions between plants and soils. Four large enclosures (35 × 10 × 3 feet) were erected and every one filled with about 50 tons of an imported soil : sand, calcareous sand, clay, and chalky clay respectively. In addition, there was *in situ* Potterne soil of Upper Greensand origin. Plants used as experimental material were all of known wild origin and had been used in experimental breeding more or less extensively so that at least something was known about their genetic constitutions. Soil was the variable factor. The beds were all near together and thus under the same climate and were so arranged and spaced that there was no shading or formation of different microclimates, i.e. climatic conditions within quite small areas. A few of the plants were tested as seedlings but the majority as clones. A clone is the offspring obtained by vegetative multiplication from one plant which started life as a normally produced seedling. With many perennial species hundreds of ramets (the physiologically independent " individuals " of a clone) can be produced within a few years. Genetically these ramets are all the same[1] and with due care in their preparation can be made practically uniform in fluctuating characters before final transplanting. At Potterne, 26 ramets were planted in every one of the 5 soils and 26 at Kew (where the climate was also different) of 11 different species and varieties altogether during the course of the experiments. Scorings (analyses and recordings of characters) were made at frequent intervals for a great range of structural and functional reactions. It was found that different species reacted very differently. A knapweed (*Centaurea nemoralis*) did well on all soils but best on sand. Most of the other species did worst on sand, which was the poorest soil in ordinary plant nutrients as shown by chemical analysis. Some species, as kidney-vetch (*Anthyllis vulneraria*) and a variety of sea-campion (*Silene maritima*) gave very different death-rates on different soils, but others did not.

[1] Except for rare body-mutations or segregations (see next chapters).

Attacks by diseases and by pests (such as slugs) varied with soils and with species. The degree and precocity of flowering and fruiting was partly correlated with soils, partly with the weather conditions (which were fully recorded) of any one year. The degree of plasticity varied greatly. The most plastic species was broad-leaved plantain (*Plantago major*), in two genetically distinct varieties. The differences in size of all organs, in the number of crowns produced, and in flowering and fruiting as between ramets on sands and those on clays was most striking. Within a year the reproductive capacity of a single ramet on clay exceeded that of all 26 ramets on sand. The development of the root systems was also different in some of the species, particularly in the bladder campions (*Silene cucubalus* and *S. maritima*). Natural seed germination showed great differences according to soil and species.

The original papers must be consulted for further details of these experiments. The point it is desired to make here is that there is a very wide scope for transplant experiments carefully carried out and fully recorded. These need not be so extensive or elaborate as those at Potterne, which were conducted at a biological station. What is essential is : (*a*) to keep well-prepared herbarium specimens of the plants as originally collected for the experiments—this enables comparisons to be made with an original standard and determinations to be checked ; (*b*) to describe and analyse as far as possible the environmental conditions, e.g. the different soils used, or sun and shade conditions, or wet and dry habitats ; (*c*) to describe structural and behaviour characters at frequent intervals for every seedling or ramet which should be numbered ; and (*d*) to devise a scheme for analysis and synthesis of the results.

Genetic Variation

We have seen that fluctuations occur in all parts of plants. It is equally true that variations with a genetic basis may also involve any organ, structurally or functionally or, perhaps always, both. In order to emphasize this ubiquity of genetic variation examples will be given under the headings of the principal organs for seed-bearing plants. Readers can add numerous examples to the few selected. Appendix F should be consulted for the significance of the few genetical terms and phrases used in this chapter.

Roots. Recent research has shown that root systems can only be studied adequately by the digging of trenches and following the branching network on an exposed face of the trench. The extent and depth to which roots penetrate is often surprising. The primary root-system of spring-sown wheat regularly reaches depths of 4 to 5 feet. In winter varieties the mature root-system has a working level of 3½ to 4 feet and a maximum depth of 5 to 7 feet. There are differences in average depths between some varieties. Root-systems have been shown to have distinctive characters in many different kinds of wild plants—grasses, trees, herbs of chalk grassland, heath species, and so on, but there is still a wide field for investigation in the underground parts of plants and we know far too little about their seasonal behaviour.

An interesting variation is shown by the lesser celandine (*Ranunculus ficaria*). The var. *bulbifera* produces small tubercles, modified roots, in the axils of stem-leaves and propagates almost entirely by these and the tuberous roots produced just underground. There are other characters correlated with the production of axillary root-tubercles, and their evolutionary significance is considered later.

Stems. Pyramidal varieties, often with almost erect lateral branches (fastigiate), are known for a number of tree species in addition to the well-known Lombardy poplar (*Populus italica*) which, by the way, is nearly always male. A collection of such, growing in the Royal Botanic Gardens, Kew, includes: alder, beech, hornbeam, oak, horse-chestnut, birch, and rowan. The habit also occurs in conifers as in the Irish yew—of which only the female was known till recently when the male was found in cultivation—junipers, and cypresses. Another variation is that of "weeping" varieties such as are found in beech, birch, and ash. Prostrate stems are characteristic of *Solanum dulcamara* var. *marinum*, a variety of bittersweet growing on shingle beaches, and of *Cytisus scoparius* var. *prostratus*, a variety of broom recorded only for a few localities near the sea. The prostrate stem-character in each of these species comes true from seed and is obviously correlated with the natural habitat conditions of the variety. The spines (modified stem-branches) of spiny rest-harrow (*Ononis spinosa*) are a contrast to the unarmed branches of the field rest-harrow (*O. arvensis*). These species would repay experimental investigation.

Foliage leaves. Many students of plant structure recognize only three main categories of organs as composing the body in the seed-plants: roots, stems, and leaves. The category of leaves is the most variable

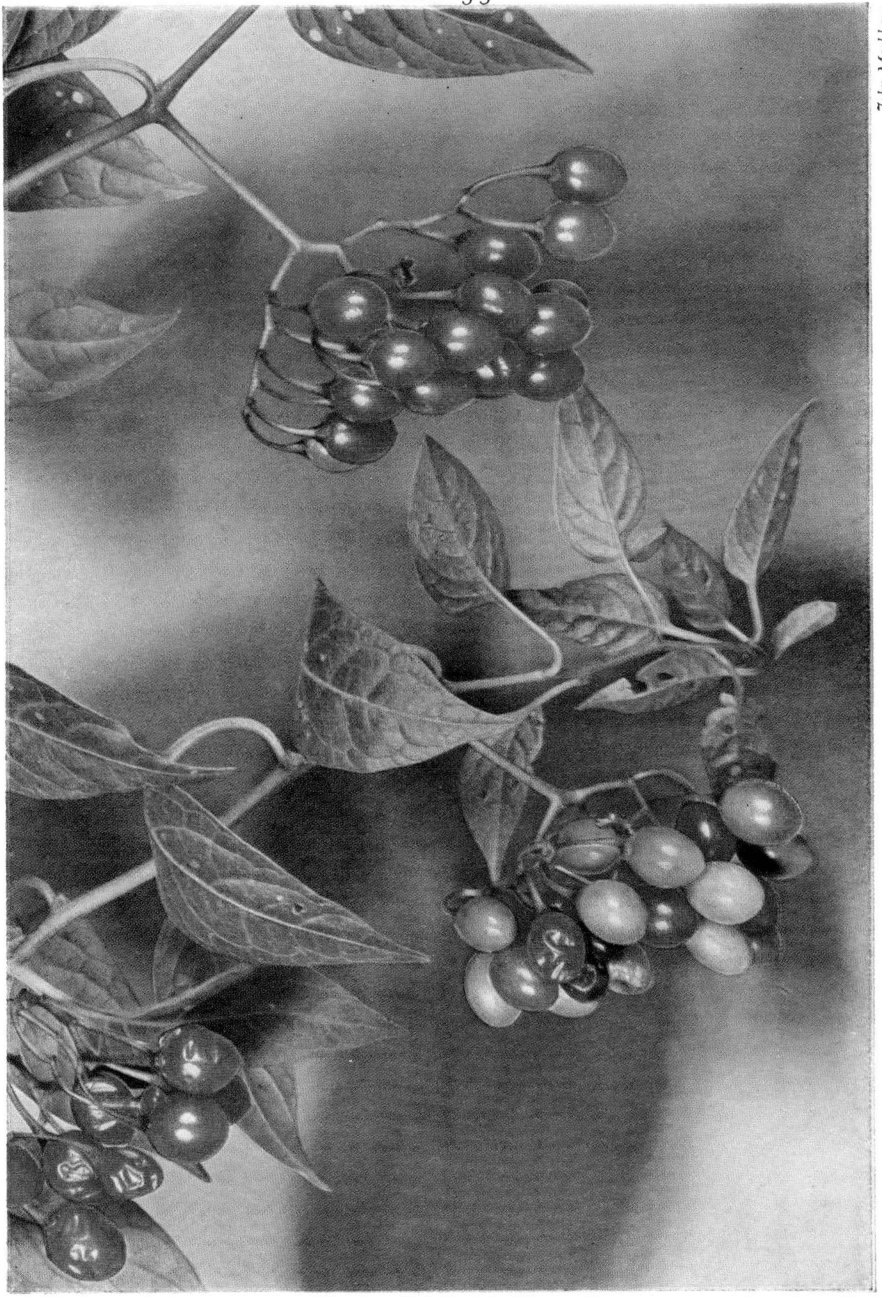

John Markham

WOODY NIGHTSHADE, *Solanum dulcamara* (Solanaceae). Shenley, Hertfordshire. September 1945

PLATE 34

F. Ballard

Oxford, April 1945

of the three in whatever sense we use the term variable. Under "leaves" are classified cotyledons, scale-leaves of various kinds (bulb-scales, rhizome-scales, bud-scales), foliage leaves and their modifications into spines or tendrils, sepals, petals, stamens, and carpels. We are not here concerned with the difficult questions of theoretical morphology and in this section will limit our illustrations of genetically fixed variations to foliage leaves.

The outline shape of a leaf is, within limits, a specific or varietal character. Some species have a large number of varieties distinguished by leaf-shape. Thus, in the common and sea bladder campions, sizes and length/breadth ratios of the leaves show a great range correlated with the genetic composition of the plants. The Oxford ragwort (*Senecio squalidus*) and the common hawthorn (*Crataegus monogyna*) are also variable in leaf-shape, but breeding work remains to be done on these plants, as it does with the British varieties of the common vetch (*Vicia sativa*). A considerable number of species have one or a few sharply marked varieties based on leaf-shape, as elder and greater celandine with cut-leafleted varieties, and stinging nettle (*Urtica dioica*) with differences in shape and margins. The creeping thistle (*Cirsium arvense*) has a number of varieties distinguished by the shape of its leaves, the degree of prickle development, and hairy covering. In goldilocks (*Ranunculus auricomus*) there are two varieties, one with deeply dissected leaves and fully developed petals and one with entire or lobed leaves and flowers without petals. Reciprocal crosses between the two varieties have shown that hybrids have intermediate petal characters but that the leaf-shape is that of the ovule parent.

Indumentum, or the hairy covering, is in some species present or absent in different varieties or is formed of hairs of more than one kind. There are varieties of heather (*Calluna vulgaris*) with different degrees of development of the indumentum. In the silver-weed (*Potentilla anserina*) there are three varieties based on the distribution of silvery hairs on the leaflets : var. *sericea* has them on both surfaces, var. *discolor* on the lower only, and var. *nuda* on neither. The fourth possibility of silvery hairs on the upper surface only does not appear to have been recorded. It should be collected and sent to Kew if found. In many species the indumentum is similar on both the leaves and the young herbaceous stems. Thus in the fleabane (*Pulicaria dysenterica*) the typical variety (Pl. 31, p. 126) is covered with a white woolly indumentum but in the var. *hubbardii* (Pl. 32, p. 127) this is absent from both stems and leaves.

Investigation is required of the indumentum variations in daisy (*Bellis perennis*) and sheep's-bit (*Jasione montana*).

Most foliage leaves are green but there are copper-coloured varieties of beech, hazel, species of *Prunus*, etc. The coloration is often most intense in the young leaves and is due to the presence of colouring matter (anthocyanin) dissolved in the cell-sap. In some herbs there are varieties with patches or blotches of colour in more or less well-defined groups of cells immediately under the epidermis (outermost cell-layer). Examples are found in buttercups and lesser celandine and in lords-and-ladies (*Arum maculatum*) (Pl. XVII, p. 226).

Inflorescences. An interesting variety of bittersweet (*Solanum dulcamara*) with one-flowered inflorescences and, later, one-fruited infructescences, was found in Northern Ireland. On being crossed with the normal (Pl. 33, p. 130) it was found to be recessive and segregated in a ratio approximating to 3 normal (*multiflorum*) to 1 one-flowered (*uniflorum*). Appendix F may be referred to in explanation of this experiment. Some grasses show a series of inflorescence variations and these require detailed investigations in such genera as *Molinia*, *Agropyron*, *Festuca*, and *Poa*. The number of flowers per inflorescence is one of the differences between the common and sea-campions, and hybrid plants have intermediate numbers. The presence or absence of ray florets in the flower heads of knapweeds of the *Centaurea nigra* group has a genetic basis.

Flowers. Variations in floral parts have naturally attracted a great deal of attention. Any or all of thc organs may be involved and the range of variation may be great or small.

Calyx variations have been recorded less frequently than those of corollas. Petaloid sepals occasionally occur, as in lesser celandine, where they are associated with maleness and absence of functioning carpels. In primrose (*Primula vulgaris*) a variety with the sepals developed as foliage leaves (phyllody of the sepals) is not uncommon. In wood anemone (*Anemone nemorosa*) varieties are known with the petaloid sepals varying in colour and shape.

In some species petal shape is very variable. In meadow saxifrage (*Saxifraga granulata*) a considerable number of genes is known to be involved in giving a wide range of shape and size to the petals. The presence of peculiar outgrowths (enations) on the upper surface of the corolla-lobes in the primrose is recessive to the usual condition of their absence. In toadflax (*Linaria vulgaris*) there is normally an irregular (zygomorphic) corolla with one spur. Varieties occasionally occur with regular (actinomorphic) corollas either with five spurs or with none at all, the condition being known as peloria. The occasional production of actinomorphic flowers also occurs in other Scrophulariaceae, Labiatae, Leguminosae, and Orchidaceae. In bluebell (*Scilla non-scripta*) (Pl. 24, p. 107) the flower-shape varies a good deal in large populations but the genetics requires investigation. Variations in garden stocks of bluebells are sometimes partly due to hybridization with another species, *S. hispanica*. In foxglove (*Digitalis purpurea*) a variety with large campanulate terminal flowers, structurally due to fusion of a number of flowers in a kind of fasciation, breeds true on self-fertilization and is recessive when crossed with the normal. Varieties in which the corolla is entirely suppressed occur in chickweed (*Stellaria media*), wallflower (*Cheiranthus cheiri*), and goldilocks (*Ranunculus auricomus*). "Double" flowers occur occasionally in many species. Usually the doubleness is a multiplication of petals, sometimes at the expense of stamens and carpels. If the doubleness be complete the variety is sterile and can only propagate vegetatively. If some stamens and carpels remain functional the genetic basis of the "sport" can be tested by crossing with the single variety. Doubleness is sometimes dominant (as in Welsh poppy, *Meconopsis cambrica*, and carnations, *Dianthus caryophyllus*), sometimes recessive (as in greater celandine, *Chelidonium majus*, and sweet william, *Dianthus barbatus*), and sometimes the F_1 is intermediate (as in hollyhock, *Althaea rosea*).

Colour in flowers is due either to pigments in special living bodies inside the cells known as plastids (flavones) or dissolved in the cell-sap (anthocyanins). The former give the various shades of yellow and the latter the reds, blues, and purples. Colour variations are mainly controlled genetically and even slight variations in shade may indicate a different genetic composition. A large number of British species normally with coloured flowers produce white-flowered varieties but with varying frequencies. In some areas of Oxfordshire and Berkshire white sweet violets (*Viola odorata* var. *dumetorum*) have increased in the last decade or two relative to the violet-coloured type. White-flowered musk-mallows (*Malva moschata*) are not uncommon. Blue, pink, and white varieties of milkwort (*Polygala vulgaris*) occur commonly. In the fritillary (*Fritillaria meleagris*) (Pls. 34, p. 131, and 35, p. 138) the colour ranges from deep blackish-purple to white. In red valerian (*Centranthus ruber*) there is not only a white-flowered variety, but a range of red-flowered plants can be seen to perfection, for example, in the railway cuttings and old chalk quarries in northern Kent. More rarely the common variety is white and the rarer variety coloured, as in the yarrow (*Achillea millefolium*), a species on which research is required. In the white campion (*Melandrium album*) and the red campion (*M. dioicum*) flower colour is a specific character and pink petals usually indicate hybrids. In comfrey (*Symphytum*), populations of plants with variously coloured flowers (pink, heliotrope, blue, etc.) may have originated through crossing between more than one species, but they require experimental study. Intensity of yellow plastid colour varies in buttercups and has there a genetic basis, as it probably has also in toadflax. In kidney-vetch (*Anthyllis vulneraria*) the inland variety is generally a shade of yellow but there is a coastal variety (var. *coccinea*) with red flowers. This hybridizes with the yellow variety to yield in subsequent generations a series of intermediate colours, since both flavone and anthocyanin pigments are involved. A large number of colour variants are recorded for the common pimpernel (*Anagallis arvensis*) (Pl. 36, p. 139) including blue varieties distinct from the blue of *Anagallis foemina*.

The term " sex " is used rather loosely in the descriptive botany of flowering plants. Strictly, the test of femaleness is the production of potentially functional female gametes or eggs and of maleness the production of male gametes or sperms. In flowering plants the female gamete is produced singly in the embryo-sac of an ovule and the male gametes (two) are formed inside a pollen-grain. The true sex organs are inside an ovule and inside a pollen-grain respectively and are microscopic structures. As a descriptive term " sex " has been applied to carpels (or ovaries) and to stamens, thence to flowers producing these organs and to plants producing the flowers. In this broader sense a plant may be unisexual (male or female) or bisexual, while a flower may be female (with carpels only), male (with stamens only), or hermaphrodite (with both stamens and carpels). A species with unisexual individuals is said to be dioecious, but if it has male and female flowers with both kinds on the same plant it is said to be monoecious. " Sex " in these senses has usually a predominantly genetical basis though there is sometimes a range of plasticity. In species normally with hermaphrodite flowers, rarer functionally female flowers are sometimes described as " male-sterile," but the simpler positive descriptive terms are used here.

Sex distribution sometimes provides differential characters for quite large groups. Willows and poplars are nearly always dioecious but occasionally individuals appear with male and female flowers in the same catkins. The annual dog's mercury (*Mercurialis annua*) is sometimes monoecious, the perennial species (*M. perennis*) is dioecious but, very rarely, monoecious individuals have been found. A rather frequent variation is gynodioecism or the occurrence, within a species, of plants with only hermaphrodite flowers and others with only female flowers. In the Labiatae gynodioecism occurs, often commonly, in many genera, as *Thymus*, *Origanum*, *Glechoma*, and *Salvia*. It has also been found in Compositae (*Centaurea*, *Taraxacum*), Caryophyllaceae (*Silene*, *Stellaria*, *Dianthus*), Saxifragaceae (*Saxifraga*), and other families. The genetical basis as worked out in some Labiatae is that two dominant genes A and B must be present to produce hermaphroditism (AABB representing a full true-breeding hermaphrodite) while with only A female plants are produced. Why males are either not produced or are exceedingly rare is not yet clear. The white bryony (*Bryonia dioica*) is normally dioecious, but plants with hermaphrodite flowers have been grown at Kew from seed of a hermaphrodite parent. There is a full range of variation in sex in buttercup (*Ranunculus acris*). Most commonly plants have hermaphrodite flowers, but individuals with only female flowers are frequent in many populations. Two fluctuating classes have been recognized between hermaphrodite and female and have been shown to be of hybrid constitution. Rarely, in wild colonies or in bred families, neuters may be found with no functional stamens or carpels and incapable of producing seeds. Only once has a male plant been found, and that in Cumberland, but using this as a pollen parent numerous male plants were obtained from seed in subsequent generations. Female flowers are relatively small and neuters very small. Male plants are rather short and generally not robust and have flowers with a larger number of narrow petals, while their leaves are peculiarly crinkled and clawed with narrower segments than usual. All these characters remained strictly associated with maleness (female sterility) in the numerous male plants derived by crossing various hermaphrodite and female plants with the male. It may well be that maleness is often, in flowering plants, associated with weak, harmful, or even lethal characters and this may account for its rarity as compared with gynodioecism.

A physiological variation may be mentioned here. It is economically important to know that some plants, notably some varieties of fruit trees such as plums, cherries, pears, and apples, are sterile with their own pollen but fertile with the pollen of some, but not all, other varieties of the same horticultural species (cultispecies). This incompatibility, as it is termed, occurs also in some herbs, as in species of speedwell (*Veronica*) and may be commoner in our British flora than we yet know. Incompatibility is due to failure of pollen-grains to function in " own " stigmatic or stylar tissues, either because the grains burst or the tubes do not grow to reach the ovules.

Season of flowering is an interesting phenomenon which shows clearly the complicated interactions between inherent constitution and external environment. Every wild species has a usual flowering behaviour but it is very probable that any plant could be made to flower, or conversely prevented from flowering, at any season by appropriately modifying certain conditions, which are known for a few species—temperature, light, and food supplies being the principal ones. In some genera (notably in *Colchicum*, *Crocus*, *Scilla*, and *Gentiana*) very similar species flower in the spring or autumn respectively. In our British flora the best example is in the red bartsia genus, *Odontites*: *O. verna* flowering in late spring and early summer and *O. serotina* in late summer and early autumn. There are slight, but usually fairly distinct, structural differences between the two microspecies, which some botanists regard as subspecies or varieties of one species. Connected with flowering is photoperiodism or the influence upon growth and behaviour of the length of daylight. Some plants will only flower when subjected to periods of light not exceeding 12 hours (short-day plants) while others must have 16 or more hours' daylight in order to be brought to the flowering stage (long-day plants). Some species are indifferent, in their flowering behaviour, to the length of daylight, as groundsel (*Senecio vulgaris*) and the Oxford ragwort (*S. squalidus*), both of which species have been found in flower in every one of twelve consecutive months.

A difference in the length of the style correlated with difference in the position of the stamens or length of the filaments gives variants in a number of species and is known as heterostyly. In primrose the short-styled (or thrum) variant is dominant over the long-styled (or pin) variant. In purple loosestrife (*Lythrum salicaria*) there are three style-lengths in different plants, which are long-, mid-, or short-styled respectively, with associated differences in heights of anthers (tristyly).

Fruits and seeds. Variation in fruits and seeds of British plants has not hitherto received as much attention as it deserves, except in a few genera. In the bladder campions there are differences in fruit shape which most often distinguish *Silene cucubalus* from *S. maritima*, but there is some overlapping of the characters concerned. In the bur-marigolds (*Bidens*) also there are differences in the fruit between *B. cernua* and *B. tripartita*. Microspecies of *Erophila* (whitlow-grass) and *Capsella* (shepherd's purse) are partly distinguished by differences in fruit shapes. Two British species of watercress are best distinguished structurally by fruit shape, seed arrangement, and seed-coat markings. In *Nasturtium officinale* the fruits are relatively thicker and shorter, the seeds are in two ranks and their surface is marked by a coarse network; in *N. uniseriatum* (*N. microphyllum* ?) they are longer and more slender, the seeds are in one row, and the seed-coat is finely reticulate. As examples of variation within one species the fruits of ash and bluebell may be mentioned. In both species there are considerable differences in shape but they need full statistical and experimental investigation. In dandelions, *Taraxacum laevigatum* has reddish fruits, while in *Taraxaca* of the *vulgaria* section the fruits are grey. Seed size, shape, markings, and colour may all show variations which are genetically controlled. The vetches (*Vicia*) would well repay investigation for seed characters (as well as for leaflet characters) in Britain. In *Silene cucubalus* white immature and tubercled mature seeds are the rule, while in *S. maritima* purplish immature and armadillo mature seeds are most frequent. White is recessive to purplish and armadillo to tubercled. In spurrey (*Spergula arvensis*) one variety (or microspecies) has smooth and another papillose seeds.

Variation in Cryptogams

Attention has been directed so far entirely to variation in the seed-bearing plants. This is partly owing to these being the dominant members of our land flora, partly to the fact that, with some exceptions, there has been more investigation into their variation than into that of the cryptogams, or plants not producing flowers and seeds, and partly to the editorial arrangements for the publication of other volumes in the New Naturalist series dealing with special cryptogamic groups. It is, however, desirable to draw attention to a few general facts regarding variation in the main groups of cryptogams since they have a bearing on problems discussed in later chapters. As we have started with the seed-bearing plants we will work downwards to groups regarded as lower in the evolutionary scale.

The largest extant group of the vascular cryptogams (those with conducting strands of wood and bast) is that of the Ferns (*Filicales*). Although without flowers or seeds, the fern plant is produced by fertilization and itself bears spores. It has a double set of chromosomes, one set from a female and one from a male gamete. It is technically termed a sporophyte and is anatomically characterized by vascular tissue (strands of conducting elements) and stomata (pores through which gases are exchanged with the outer atmosphere) and structurally by the possession of roots. Variation in roots, stems, and leaves (often called fronds because they bear the spore-bearing structures) is thus comparable genetically to that in the vegetative parts of seed-bearing plants. In the superficial descriptive sense variation has been extensively studied in ferns, and almost innumerable "varieties" have been named from differences in the size and shape of the fronds. In a few species these variations have been shown experimentally to have a genetic basis. It must also be noted that in some species there are great differences between the fronds on one plant, and these differences are sometimes correlated with the age, function, and position of the fronds. Little is known regarding variation in the other British vascular cryptogams: horsetails (*Equisetum*), club mosses (*Lycopodium* and *Selaginella*), and quillworts (*Isoetes*).

The mosses and liverworts offer an exceptional field for study of variation because the vegetative plant gives rise directly to the gametes and normally has a single set of chromosomes, i.e. it is a gametophyte. Moreover, owing to the ease with which vegetative multiplication can be induced in mosses from pieces of the capsule or seta (the sporophytic parts of the moss, with the double chromosome number) without reduction of chromosome number it is possible to obtain gametophytes of pure or hybrid origin with two or more times the normal number of chromosomes and to compare these with ordinary gametophytes. Apart from such fascinating though somewhat difficult research there is much to be learnt by comparative studies of variants within widely ranging species. Thus, *Hypnum cupressiforme* is a common moss occurring on semi-stabilized sand-dunes, on heaths, on walls and rocks, on tree-trunks, and in other habitats. A large number of varieties and forms of this species have been described but their value and relationships are

F. Ballard

FRITILLARY, *Fritillaria meleagris* (Liliaceae). Magdalen Meadows, Oxford. April 1945

PLATE 36

Brian Perkins

SCARLET PIMPERNEL, *Anagallis arvensis* (Primulaceae). Near Leatherhead, Surrey. September 1946

only imperfectly understood. A study, by modern methods, of variation in such a moss comparable to studies made in some flowering plants and recorded in the following chapters would be most instructive.

It is scarcely possible to generalize about variation in Fungi and Algae because these groups are so large and so varied in size, structure, duration, life-histories, and habitats. Naturalists with a flair for experiments on microscopic objects might well undertake research on the plasticity of freshwater Algae since their environmental conditions can be controlled very efficiently and with relative ease once their general life-history is known. Again, there is an open field for research on variation in seaweeds for anyone living near the coast. Variation in microscopic Fungi and in Bacteria can only be studied adequately by laboratory techniques. Work that has been done on these structurally simple thallophytes (plants with no differentiation into root, stem and leaf) shows that they are often highly plastic and that marked variations can be impressed by modification of environmental conditions and that the resulting "saltations," as they have been called, are sometimes of long duration.

Special Kinds of Variation

Hybridization between species or subspecies often increases variation by making possible more gene-combinations and by making some of the hybrids more plastic than are non-hybrids. This recombination variability may be great in hybrid swarms where, in addition to first crosses, there are all sorts of back-crosses with the original parents and between segregates. Excellent examples of hybrid swarms have been recorded in this country in field pansies (*Viola*), knapweeds (*Centaurea*), wood avens (*Geum*), hawthorns (*Crataegus*), comfreys (*Symphytum*), and marsh orchids (*Orchis*).

Somatic, or body, variation of a genetic kind is of rare occurrence in wild plants. It may be due either to a mutation (a change in a unit or a mechanism of inheritance) occurring in a body cell whose descendants come to form a growing-point, or to some form of segregation taking place in the body tissues of a hybrid—the cells with the segregated genes likewise becoming the growing point of a branch. Segregation here means the separation of units of inheritance brought together by previous hybridization. A striking example has been recorded in the pimpernel (*Anagallis arvensis*) with a portion of the plant producing scarlet and the rest of the plant blue flowers. Different colour variants occasionally occur on the same plant of some heaths, as *Erica vagans* and *E. ciliaris*.

One rather difficult question should be mentioned here. Do variations sometimes represent reversions to an earlier ancestral form ? The possibility of this cannot be denied but such an interpretation must be made with caution. Reverse mutations have been obtained and presumably one may then represent a more original condition than the other. Both fluctuations and sports are sometimes supposed to throw light on the structural nature of organs. A spine may, as in gorse, be fundamentally of stem (branch) nature and may be described as a metamorphosed stem. If conditions be varied so that it grows out into a leaf-bearing stem this may, in one sense, be described as a reversion. A sepal may be a metamorphosed foliage leaf. In goldilocks a sepal may be varied to foliage-leaf size, shape, and function, but it may also develop as a petal. Both cannot be reversions and it is doubtful if the word be applicable to either. Certainly reversion is no mysterious phenomenon of return to a far-off ancestral condition or atavism.

Functional or physiological characters usually require more technique for their study than do structural or morphological characters, and though their importance is obviously great, less is known about their variation. The vitamin C content of rose hips and of apples varies both with environment and with the species or variety. The bird's-foot trefoil (*Lotus corniculatus*) has two varieties, morphologically indistinguishable from one another it is said, the one yielding hydrocyanic (prussic) acid and the other not.

Concluding Remarks on Variation

We have used the term " variation " in this chapter in a very wide sense. An adequate classification of the different kinds of variants has not yet been published, and since it would have to include variation in plants of all countries and in cultivated kinds (cultispecies) as well as in those of the British flora it cannot be attempted here. It is, however, evident that a great deal of research is needed on this subject within the realm of our own flora and especially in common and widely ranging species. Such research must involve not only the recording and describing of strikingly aberrant variants (" sports " or " teratological variants ") which appear very occasionally, but also the different variants which may be found in considerable numbers, more or less constantly and regularly, in this or that or in every population of large size in many, if not all, species. Such studies on wild populations must be made with the help of simple statistics. If repeated over a period of years they may well throw light not only on the constitution but also on the evolution of species. The proper study of variation also demands experimental treatment, though again it is clear planning and care in simple details rather than elaborate schemes and expensive apparatus that are essential. It is an interesting fact, true in many branches of study other than botany, that the laws governing the normal can often only be investigated through contrast with the abnormal, or perhaps one should say more generally, opposites must be found before discoveries can be made. The naturalist has to use two apparently contradictory methods: to note every variation independently and to compare and contrast it with the normal, the mean, or a standard (a famous geneticist said " treasure your exceptions ") ; and yet at times he must be prepared to generalize, and this means either ignoring exceptions or merging all exceptions in calculations giving some mean value, which may be entirely artificial in the sense that it has not been found in any organism. In the next and subsequent chapters we shall see how the use of these different methods throws light on the composition, behaviour, and origin of British plants.

CHAPTER II

ADAPTATION AND NATURAL SELECTION

ADAPTATION is another of the words taken over by biologists from everyday language and given a special or at least a more restricted and precise meaning. It is obvious that every species is adapted to its habitat range to the extent that it can survive within it, such species-survival involving not only establishment and growth of individuals but also reproduction. In addition to this truism there are numerous examples of plants with special structures or functions which make them peculiarly suited for living in special environments or for taking advantage of special features in an environment. None can dispute that adaptation is one of the most striking phenomena in plant life and one to which the attention of the field naturalist is everywhere drawn. There are, however, two negative generalizations to be recognized. The first is that because a plant is adapted to a given environment it must not be concluded that the plant, or its ancestors, in any sense deliberately adapted form and function to that habitat. Some accounts of adaptation give the impression that a plant sits down and considers how it can adapt itself to some new or special condition in order to survive and compete successfully. This is just nonsense. A plant has to be studied scientifically as being completely under the dual control of heredity and environment, since there is no evidence that it can take "thought for the morrow." The second is that it is not a valid deduction, from the fact that adaptation is widespread, that every character has some special adaptational value. As we shall see, the concept of adaptation is complicated and we must be prepared to acknowledge the occurrence of characters of all degrees of direct and indirect value to the plant showing them. There are many characters which, so far as present evidence allows of a reasonable conclusion, appear to confer no particular advantage on the plants possessing them. We must be very careful in either suggesting or denying adaptation without good observational or experimental evidence.

If by adaptation, then, we here mean that a plant is suited by its structures and functions to grow and propagate under a given range

PLATE V

John Markham

East Barnet valley, Hertfordshire, showing hoar-frost on birch and elm trees. January 1944

PLATE VI

of environmental conditions, we have still to remember the complexity of the environment. In an earlier chapter it was pointed out that a plant's actual environment can be analysed under the main headings of climatic, edaphic (soil), and biotic factors. Many plants, especially herbs, are restricted to certain microclimates ; for example the light, temperature, moisture, and other conditions within a wood are markedly different from those in open ground under the same general climate, and we have well-adapted woodland plants in our British flora. For " soil " factors it would often be better to substitute " substratum " or " terrain " factors, for plants frequently grow in or on what the farmer or gardener would certainly not accept as soil. Biotic factors are still more diversified, including as they do all the relationships between plants and animals and between plants and plants.

In the previous chapter on variation we have seen that plants may vary in any of their organs and in many different ways. We distinguished fluctuating from genetic variation but did not emphasize in our analysis the criterion of survival value. In the next chapter we shall deal with the cytogenetics of British plants and see how complicated inheritance may be. We have here to consider this major question : given abundant variation, can we account for the undoubted fact that plants are adapted, more or less strikingly, to this or that environment —to a woodland, a heath, a bog, a salt-marsh, a rock, etc. ? Three theories have been held, with various modifications. Firstly, that the different species of plants were created once and for all independently one of another, with their existing characters, that they did not arise from pre-existing species, and do not give rise to new species. This theory is now disproved by our knowledge of the fossil record and by experimental and observational evidence. Secondly, it was postulated that fluctuations became " fixed " ; in some way modifications induced in the organs and functions of a plant were transmitted to subsequent generations. Since, however, only the sex cells (gametes) are transmitted from generation to generation it can only be by changes in their constitution that new characters of specific or varietal value can be induced and maintained in subsequent generations. The exact causes of mutations (changes in genes—whether in chemical or physical structure, number, or arrangement) are still obscure, especially for natural environments, but the weight of evidence is against the view that externally induced bodily changes are faithfully reflected in some kind of parallel change in gene constitution. That environmental factors can induce gene and chromosome changes is proved by experiments with X-rays, radium emanations, chemical substances (e.g. colchicine), high and low temperature, etc., but these changes are not adaptational (except by a possible rare chance) and are not parallel to previously induced bodily changes (if any). Mutation, in fact, whether occurring naturally or induced artificially, appears to be most often entirely at random relative to adaptation to features of the surroundings. Adaptation, however, is not random, it is obviously directed by the environment interacting with the genetic constitution. We thus come to the third theory, that adaptation results from the natural selection of variants showing characters of form and function best enabling the species to grow and reproduce under the given environmental conditions. We shall see how far this theory, which is now generally held by evolutionary biologists, is supported by research on British plant life.

In our accounts of habitats and of plant communities, we noted a general absence of great natural extremes in the British Isles. Here are no deserts, no permanent snowfields, and no great mountains. Our climate is temperate and oceanic, and variations from north to south or from east to west are neither so great nor so abrupt as in many other countries. Trees are naturally dominant over the greater part of the surface and the climax vegetation is forest. In soils, it is true, there is considerable diversity, but, again, we have no inland "bad lands" of alkaline soil, no extensive tracts of loess or laterite—our soil maps are patch-works and "micro-edaphic" factors are here probably as important from the standpoint of adaptation as differences between the major soil classes. Biotic factors are always with us, and of these the competition between plant and plant is of great importance, but apart from the influence of man, there is nothing extreme in their action—we have no great herds of wild herbivores roaming unrestricted throughout the land—and, in a sense, natural (i.e. non-human) biotic factors are in the British Isles "micro-biotic." On the other hand, this general normality of so much of our country offers exceptional opportunities for a study of adaptation. One is neither distracted by the abnormal and grossly exceptional nor bored by the monotony of uniform extremes. The micro-climatic, micro-edaphic, and micro-biotic factors vary within small areas and comparisons are thus easily made *in situ* and under field conditions, while comparative observations and experiments can be made to merge or at least can be carried on together, and all without waste of time in traversing great distances. Moreover, there are exceptions to the generalization that the environmental conditions are not extreme in the British Isles. We have sufficient unusual habitats for comparison with those carrying a " mesophytic " plant life, that is, one not growing in extreme habitats. In particular, we can draw on inland waters (rivers, lakes, fens, etc.) and on coastal communities (sand-dunes and salt-marshes) for examples.

It is possible to arrange examples of adaptation in any one of several different ways : systematically, by the organs affected, or on an ecological basis. Since adaptation is suitability to environment, the last would seem to be most satisfactory. It has the further advantage of making us realize that natural selection acts through increased survival-chances for whole plants, and through them for species, and not merely for organs. It is important always to keep in mind that survival of species is dependent on continuity of " success " throughout the life-history of a plant and that this must be maintained generation after generation. To avoid repetition, however, and at the same time to emphasize some specially important aspects of ecological study, we may partially separate off the general vegetative life-history from the reproductive phases, considering the former to begin and the latter to end with establishment of seedling or sporeling. For the average seed-bearing plant, the life-history has ecologically two short crises separated by periods of longer or shorter duration when the struggle for existence is less acute, though it may be more prolonged. This gives the following scheme :

period of vegetative growth ⟶
⟶ flower production and pollination ⟶
⟶ seed-setting and dormancy ⟶
⟶ seed dispersal leading to ⟶
⟶ germination and seedling establishment.

Pollination and germination (with establishment of seedlings) are the crises. The study of actual seed formation, including fertilization and the formation of an embryo, involves the use of the compound microscope, and this is true also of the reduction division which precedes the formation of sex cells and, usually, pollination. Little is known regarding the influence of the natural environment on these phases of the life-history.

Adaptation in General Life-History, with Special Reference to the Period of Vegetative Growth

Halophytes.

Plants living in a medium or on a substratum rich in common salt (sodium chloride) are termed halophytes. An arbitrary figure of 0·5 per cent common salt has been taken as the measure for saline soils having a halophytic vegetation. In this country halophytes are almost limited to the sea (submerged halophytes like seaweeds, marine Algae, and sea-grasses, *Zostera* spp.) and to salt-marshes. The submerged halophytes show structural adaptations in their vegetative parts more to submergence in a watery environment than to the salinity of the medium. Their adaptation to the salt content of sea-water (3–3·3 per cent of salt) is mainly internal in that the protoplasm (living material) of the cells is neither killed nor prevented from functioning by the salinity of the surroundings or by any higher concentrations of salt in the cell sap. This internal adaptation to high salinity is, indeed, a general characteristic of halophytes.

Salt-marsh plants are very varied. Not only are quite a number of different families represented but the habit is by no means uniform. While some structural adaptations can be suggested it is probable that physiological adaptations are much more important. Unfortunately, these latter are difficult to study under natural conditions and, to put it mildly, published accounts of researches are not always easy to correlate one with another. Sea-water is saline mainly because of its content of common salt, but it contains other salts besides, and they may be much more important, even in small quantities, than is at present realized. Further, conditions on a salt-marsh are not uniform and evidence is accumulating that some species are adapted to conditions slightly different from those in which others flourish. The majority of salt-marsh halophytes are perennial. The common glasswort (*Salicornia herbacea*) and sea-blite (*Suaeda maritima*) are exceptions and these are early colonizers on mud. A number of salt-marsh plants are succulent, that is their leaves, and sometimes stems, have many cells filled with a juicy sap. Salt, and more especially the chlorine part (or ion) of it has been shown to induce or increase succulence in quite a number of different species of plants, and it remains doubtful if succulence gives in itself any advantage to such salt-marsh plants as show it. There is, in general, no diminution of transpiration (loss of water vapour) in salt-marsh species relative to that of plants growing in non-saline habitats. Absorption, however, presents some difficulties and the high salt content of the sap inside the plant is probably of importance in increasing the power of absorption from the saline marsh solution. It has been shown that at least some halophytes vary their absorptive mechanism according to the salt-content of the medium around their roots at any one time.

A majority of our British halophytes can flourish in non-saline soil, as in a garden or experimental ground, so long as they are not subjected to intense competition from other more quickly growing or taller plants. Sometimes they are slow to flower under these conditions, but some, as sea plantain (*Plantago maritima*), flower and fruit well under cultivation in ordinary garden soil. This suggests that salt-marsh plants naturally grow in their particular habitat because their protoplasm is not harmed by high salt concentrations while they are not subjected to competition by non-halophytes, which are kept out of saline habitats by having protoplasm which is harmed by high salt concentration, probably mainly by the sodium part of it. Halophytes also have the power of absorbing water from saline solutions. Few studies have been made of the reproductive processes in salt-marsh plants. Pollination needs investigating in them and also seedling establishment. Dispersal of seeds (or fruits) is mainly in sea-water by tidal drift and currents. There are a few published observations which suggest that the seedlings of some salt-marsh plants, unlike those of non-halophytes, have powers of establishing themselves on periodically flooded surfaces with high salt-content. This may prove to be one of the most important of their adaptations, but much further evidence is required. The details of the vegetative multiplication of some of them also require more investigation.

Plants of Sand-dunes.

We have no climatic deserts in the British Isles and few species markedly adapted to dry habitats. The nearest natural approach to sandy desert vegetation we have is found on the hills and higher slopes of young sand-dunes. The loose, permeable, mobile sand is poor in humus, and water is a limiting factor, though not the only one, to the establishment of many species. Three kinds of adaptation are particularly striking. Firstly, there are the ephemerals which flower, fruit, and die in the spring and early summer. These are not restricted to sand-dunes but are a striking feature of them and like the "acheb" of the Sahara, are adapted to dry conditions by surviving through the greater part of the year (and especially the hottest and driest part) in the form of seeds which, with few exceptions, are the most resistant to drought of all phases of a seed-plant's life-history. Examples are mouse-eared chickweeds, whitlow grass, wall or rue-leaved saxifrage, and early scorpion-grass. Secondly, there are certain biennials and perennials with tap-roots which penetrate deeply into the sand and thus enable the absorptive system to reach the permanent water-table. Ragwort, hound's-tongue, and dandelions are good examples. Thirdly, there is a small group of true sand-dune species, which tend to be the first colonizers of newly formed dunes and are almost or quite restricted to dunes. The most important of these is the marram-grass. It shows adaptation to the sand-dune environment in its anatomy, external structure, and behaviour. In particular, it has a very extensive under-sand development of buried stems (rhizomes) and roots which extend deeply and widely, but even more important is its power of growing through the sand which the wind constantly piles up over its growing shoots, new rhizome-root systems being continually developed from the buried stems.

Aquatic Plants.

A water habitat is in many ways the opposite of a dry-land habitat, such as that of sand-dunes, and provides one of the most extreme habitats for British plant life. Further, there are few districts without rivers, streams, lakes, or ponds where the many problems of aquatics can be studied at first hand. The structural characteristics of aquatics are well known and many of them are described as adaptations, though available experimental data sometimes make it doubtful whether the nature of the adaptation is yet fully understood. Amongst common structural (including anatomical) features, to which there are some exceptions, are : poor development of the root system, reduced conducting and mechanical tissues with the consequent occurrence of weak, though sometimes long, stems and flaccid leaves, finely cut submerged leaves and entire or slightly lobed floating leaves on often long and elastic leaf stalks, the development of very large air spaces which provide an internal atmosphere at least making the plant buoyant, the absence of a cuticle (thickening to outer cell walls) and functional breathing pores in the submerged parts, the production of brood-buds or other means of vegetative multiplication, and perennial habit. It is very instructive to compare the structure of organs submerged at different depths in the water with those rising above the level of the water in the same individual plant, as in mare's tail and water-violet. Such a comparison often suggests that the term "reaction," rather than "adaptation," of the plant to its environment more truthfully describes what is seen. Further, much can be learnt by studying the plasticity of aquatics as shown by a series of specimens of one species growing in deeper water, in shallower water, on mud, and on drier soil, as in floating persicaria, starworts, and water-buttercups. Nevertheless, the facts remain that aquatic vegetation is limited in the number of species of which it is composed and most of these are found *only* in water; that these species belong to a number of families often not closely related one to another but show many similar structural peculiarities; that they often have very wide distributions; and that they have to compete only amongst themselves and not with land plants. It is probable that many if not all of the aquatic seed-plants have been derived (but not always immediately) from land ancestors. The suggestion that land plants took to an aquatic environment for the same reason that Scotsmen come to England—because competition is less keen—at least needs the corollary that by mutational adaptability and plasticity they changed and were able to survive in their new homes. Actually the degree of competition in aquatic environments varies very greatly. Often fewer species are involved than on the land; the differential environmental factors (e.g. depth of water) are markedly zoned; as a result the struggle for existence is largely with other individuals of the same kind, when there is general suitability to the restricted environment. We have also to recall what we learnt in the chapter on plant communities—that freshwater vegetation represents one or another relatively early stage in the succession of vegetation, and sooner or later, and more or less gradually, the vegetation of any one stage is subject to competition by plants of the next stage, for which it largely prepares the way. Successive stages, frequently represented by zones, are most often dominated by plants which are widely separated in a general system of classification, and this suggests that their adaptations are not the result of any direct environmental action, otherwise zones of closely related species or varieties would be expected. Marked adaptation usually means extreme specialization together with loss of further or potential adaptability. Some striking examples of parallelism, in characters adaptational to aquatic life in species and genera not at all closely related, occur in British plants. One example is the resemblance in leaf characters of the water lilies *Nuphar* and *Nymphaea* (family Nymphaeaceae) and *Nymphoides peltatum* (Gentianaceae).

Calcicoles and Calcifuges (plants flourishing with and without lime respectively).

In the British Isles soils are very largely immature and consequently their characters are often closely correlated with the nature of the parent rocks. Our strata outcrop in such variety of geological age and rock constitution that our soils have a more patch-work pattern than in many other countries. Careful study of the flora and vegetation shows that these vary considerably with the type of soil, even using this term in a restricted sense to exclude such substrata as those of salt-marshes and sand-dunes which are still "soils in the making." Here we will take only the contrasts between plants on calcareous soils, such as those directly on and derived from chalk or oolitic limestones, and those on non-calcareous soils, such as those derived from Bagshot Sands or Lower Greensand. The former are termed calcicoles and the latter calcifuges. Few more striking examples of floristic and vegetational differences based on soils can be seen than those of the Chalk slopes of Box Hill and the Lower Greensand of Leith Hill though they are only a short distance apart to the north-west and south-west of Dorking respectively. There are many other examples also worth careful comparison. All limy soils are not the same physically or chemically and the same is true for non-calcareous soils. A careful comparison is still required between the vegetation and flora of our Chalk Downs and those of our oolitic and other limestones. Sands and clays may both be deficient in lime but their other characters, and their plant cover, may be decidedly different. Markedly calcareous soils react selectively on vegetation through their alkaline reaction, which neutralizes the acid products of decay and favours the development of Bacteria, especially of nitrogen-fixing Bacteria, and of the soil fauna, including earthworms. It is probable that lime, under some conditions, makes absorption of potash more difficult and thus tends to eliminate species requiring much potash. Physically chalk and limestones give dry soils—they are porous and have a high absorptive capacity but are well aerated. The abundant bacterial and animal life retards accumulation of humus through rapid oxidation. The chemical and physical features naturally interact, but broadly speaking a group of calcicoles for which the physical features of the habitat are the more important can be distinguished from another group for whose existence the chemical factors appear to be decisive. In this country the beech, common juniper, box (*Buxus sempervirens*) (Pl. 37, p. 178), lucerne (*Medicago sativa*), old man's beard (*Clematis vitalba*), and wall germander (*Teucrium chamaedrys*) are examples of the former group. Of the second group mention may be made of maple, dogwood, milkvetch (*Astragalus glycyphyllos*) (Pl. II, p. 3), dog's mercury, common buckthorn (*Rhamnus cathartica*), and ramsons (*Allium ursinum*).

The problems of adaptation to soils are even more complicated than already suggested. Many plants that are calcicolous in the wild will grow in garden soil deficient in lime so long as they are not subject to competition ; and some calcifuges will similarly grow in lime-rich soils. In other words, it is not merely a question of tolerance but of degree of reaction relative to that shown by other species which reach the same habitat. Competition is often keenest, at least on the basis of numbers involved, in the early stages of growth soon after seed germination. An excellent example of competition has been investigated in the bedstraws. *Galium saxatile* normally grows on acid soils (heaths, etc.) and *G. silvestre* on soils rich in lime. Each, however, when alone will grow well on either calcareous or non-calcareous soils, but when grown in mixed cultures *G. saxatile* kills out *G. silvestre* on acid soils, and *G. silvestre* overwhelms *G. saxatile* on calcareous soils. Structurally the two species are not greatly different and the adaptation must be physiological rather than morphological.

Bog vegetation gives an extreme example of calcifuge plants. Comparison of the plant life of bogs with that of chalk grassland shows that the two habitats have no species in common. The adaptations to the different physical and chemical conditions are both functional and structural. In particular, bog plants grow in water-logged soil but have to absorb their water and raw food materials from a highly acid medium poor in soluble salts. Only internal adaptations, not yet fully understood, enable plants to survive in such a habitat and the species occurring are relatively few in number and sharply restricted in natural habitat-range. Structurally the plants are mostly low-growing evergreen perennials with small leaves. They are frequently associated with fungal partners in their roots.

Woodlands.

Plants growing in woods have protection from wind and to a certain extent from temperature extremes. They have a humus-rich soil and are not likely to suffer from either a deficiency in or an excess of soil water. Light, however, is the main limiting factor and we find interesting adaptations of plants in the lower strata of woodland vegetation to light and shade conditions. Where a very dense shade is cast, as for example in evergreen yew woods or dense beech woods, there is practically no vegetation—except plants growing on dead organic remains (saprophytes) and sometimes a few shade-tolerant mosses. In deciduous woods (i.e. those losing their leaves in the autumn) with less dense canopies, the degree of light penetrating at different seasons of the year largely determines the extent and nature of the vegetation and the adaptations shown by its constituent plants. The period of the year when trees and shrubs are devoid of foliage is referred to as the "light phase," and the period when the foliage of the upper layers intercepts much of the light is known as the "shade phase." The former coincides with the period of lower temperatures. Carbon dioxide concentration may be appreciably higher in a woodland atmosphere than outside. Light, temperature, and carbon dioxide concentration all affect carbon assimilation or photosynthesis, that is the manufacture of sugars and other carbohydrates by the chlorophyll. Light, in general, has a retarding effect on growth in length and (within limits) on size ; in the shade, therefore, plants tend to have elongated stems and large leaves.

It is well known that soil characteristics are often very important in determining the distribution of a species near the limit of its climatic range. The selective influence of soils is well seen in the distribution of the different kinds of British trees, many of which are near the northern limits of their ranges (see pp. 105 *sqq.*). The shrub and herb layers in British woodlands tend to be dominated, in any one place, by one or a few species forming well-marked tiers. Thus there may be a carpet of bluebells on the ground with a layer of hazel shrubs above. Most of our woodlands are bare of leaves from late autumn to the spring, and we find amongst woodland plants a number of different adaptations to this factor of their environment. These adaptations are the more striking because of the uniform layer of plants showing them. One of the most obvious adaptations is the succession in time of leaf development corresponding to the relative position of the different layers. The average dates for the development of new leaves by woodland herbs which lose their leaves in winter is 19th February, for members of the shrub layer it is 21st March, while the full tree canopy is not attained till about 13th May.

However, members of the herbaceous layer in different woodlands do not all behave alike in their production and utilization of foliage. From this point of view four main types have been recognized amongst British woodland herbs. These types are :

1. *Pre-vernal type.* The foliage is formed in the early spring or even in the previous autumn and withers soon after the complete development of the shrub and tree canopy. These plants make food only during the light-phase and a striking adaptation to deciduous woodlands is their early formation of new assimilating foliage before the trees and shrubs develop a canopy which cuts off much of the light. Examples are : lesser celandine, moschatel (*Adoxa moschatellina*), lords-and-ladies, wood anemone, ramsons, and bluebell.

2. *Summergreen type.* The foliage persists throughout the greater part of the shade phase. The leafy shoots of dog's mercury first appear in January but the fronds of bracken not till May. Both these plants flourish in moderate shade. Adaptation must be of an internal kind, in so far as sufficient food can be manufactured under conditions of reduced illumination for such vigorous growth that dense layers of one species are formed which effectively prevent the entry of other species.

3. *Wintergreen type.* New leaves are formed in the autumn but some or most of the foliage may persist throughout the year. Examples are creeping bugle (*Ajuga reptans*), ground-ivy (*Glechoma hederacea*), wood-sorrel (*Oxalis acetosella*), wood-sanicle (*Sanicula europaea*), and archangel (*Galeobdolon luteum*).

4. *Evergreen type.* The green leaves persist throughout the year as a full complement of foliage. Examples are periwinkle (*Vinca minor*), fetid iris (*Iris foetidissima*), fetid hellebore (*Helleborus foetidus*), and ivy. Butcher's broom (*Ruscus aculeatus*) belongs to the same type, but here the green leaf-like organs are flattened stems.

In underground parts also, woodland plants are adapted to the peculiarities of the environment. Plants with creeping underground stems (rhizomes) are very common in woodlands. These rhizomes are obviously very suitable for penetrating the loose layers of leaf mould and enabling the plant to spread very efficiently, even into territory already occupied by other plants in which seedlings could not establish themselves. The field layer in woods is mainly composed of herbaceous perennials and these show interesting adaptations in the mode of emergence of the vegetative shoots. Three main kinds of emergence have been recognized. Firstly, that with a crown at the surface of the soil, the new leaves easily pushing up as erect structures through any covering material, as in primrose and foxglove (*Digitalis purpurea*). Secondly, that of spear shoots which are pointed and more or less protected by young leaves or scale-leaves, as in ground-ivy, speedwells, rose-bay willow-herb (*Epilobium angustifolium*), lesser celandine, bluebell, and orchids. Thirdly, that of the bent shoot or bent leaf-stalk by means of which the developing organ emerges without injury and then, by reaction to light, straightens itself out, as in dog's mercury, wood anemone, wood sanicle, and winter aconite (*Eranthis hyemalis*).

One very specialized, though small, group of plants is that of the saprophytes. Many Fungi and Bacteria belong to this group and are abundant in woodlands. Since they do not manufacture their own carbohydrate food and, indeed, have no chlorophyll by which to do so, they are to that extent independent of light. The woodland shade is not a limiting factor to their existence, while the rich forest humus of decaying leaves and twigs provides them with an abundance of already manufactured food. It is probable that many seed-bearing plants can use organic materials occurring in the soil before they have been completely broken down to relatively simple inorganic compounds. We have, however, few complete saprophytes in the British flora. The two best examples are both woodland plants: the yellow bird's-nest (*Monotropa hypopithys*), which is often placed in the heath family (Ericaceae), and the bird's-nest orchis (*Neottia nidus-avis*) of the orchid family (Orchidaceae). In some conspicuous features these two plants are much alike—a good example of the occurrence of superficially similar characters, especially vegetative ones, in plants which belong to widely separated systematic groups. The reproductive parts are quite different in essential structure. Both plants are perennial herbs of a yellowish-brown or brown colour, with no green chlorophyll. The leaves are represented by scales arising on the flowering stems which grow erect and have rather densely arranged flowers of much the same colour as the stems and scale-leaves.

Mycorrhizal Plants.

Somewhat intermediate between saprophytes and parasites (organisms living on and obtaining their food directly from other living organisms or hosts) are those plants which have a fungal partner associated with the food-absorbing organs (these are the roots in most seed-bearing plants). The combination of a Fungus and a higher plant-organ is termed a mycorrhiza. A very large number of species have mycorrhizas. They occur in liverworts, in ferns, in gymnosperms, and in many families of the flowering plants. They are recorded among herbs, shrubs, and trees. The phenomenon is usually regarded as giving reciprocal advantages to the two partners. There is a good deal of evidence for this ; though a mutually helpful living together, or symbiosis, may be a more accurate description for some mycorrhizas than for others. It is possible that some mycorrhizas began by a Fungus parasite attacking a higher plant, but the latter kept the parasite in check and ended by utilizing it. Be that as it may, the facts remain that mycorrhizas are of widespread occurrence and that the plants having them flourish, while if the mycorrhizas do not develop the plants make poor growth and even die. Mycorrhizas are very varied in their size, shape, structures, and detailed behaviour, and generalizations regarding their adaptational value to one or both components must at present be accepted with some caution. The Fungus component (and a number of species of Fungi have now been identified as connected with mycorrhizas) is unable to manufacture food for itself and, if completely enclosed in the cells of the higher plant, must depend on the latter for its food. When the Fungus is thus internal to the other component (endotrophic mycorrhiza) it probably helps in some form of digestion of organic foods taken up from the soil. When the Fungus is (at least largely) outside the higher plant (ectotrophic mycorrhiza), usually surrounding rootlets, it may help in absorption as well as in the preliminary digestion of organic substances ; while in its turn it may benefit by receiving some utilizable food materials from the higher plant.

Many forest trees have mycorrhizas, usually on superficial root systems. A good example can be seen in the beech. If the superficial layers of fallen leaves under beech trees be carefully brushed away an extensive, much-branching system of beech roots will be found only an inch or two below the surface. On these roots are numerous thickened rootlets of a pale brownish-yellow colour. Microscopic examination shows that these are mycorrhizas. If collected they should either be packed with moist leaves in a tin or placed in fixing solution in a tube with as little delay as possible since they quickly dry up on full exposure to the air. Where the leaf-mould layer has been removed the development of mycorrhizas is much reduced and the trees may then show poor growth. This is also true of other tree species besides the beech.

The story for the heather (*Calluna vulgaris*) (Pl. 17, p. 90) is even more complicated, since its mycorrhizas develop naturally only when the soil is acid and bacterial growth prevented or greatly reduced.

Most or all orchids are also mycorrhizal plants. Under natural conditions the appropriate Fungus partner must be present in the soil and enter the root of the very young seedling before further growth takes place. The seeds of some species of orchids can be made to germinate and the seedlings to develop successfully by growing them on sterilized media containing sugars and other chemical substances under aseptic conditions. This suggests that in nature the Fungus partner has the power, probably by the activity of ferments (enzymes) which it produces, of providing suitable organic compounds (sugars, etc.) from the humus of the soil and that without these substances successful seedling development cannot occur.

Carnivorous Plants.

A limited number of plants, in the British flora, mostly growing in bogs, boggy places, or acid water, supplement their food supply by capturing and digesting the bodies of small invertebrates (often insects ; hence they are sometimes called insectivorous plants). It is probable that nitrogenous salts are difficult to obtain from boggy, acid water and that it is nitrogenous compounds that are the most needed materials obtained from the bodies of captured prey. There is no doubt of the structural adaptations for the capture of small animals in these carnivorous plants. We may illustrate them by reference to two examples in which the adaptations are markedly different though the result is similar.

The common sundew (*Drosera rotundifolia*) is frequent in bogs where it is often associated with bog mosses (*Sphagnum* spp.). It is a low-growing plant with a basal rosette of leaves whose upper surfaces entice and entrap small animals. On the surface and from the margins of the rounded blades there grow numerous stalked and relatively stout hairs, every one of which ends in a rounded-ellipsoid gland surrounded by a viscous fluid. The leaf, which contains green chlorophyll, is suffused with a red colour dissolved in the sap (anthocyanin).

The colour and the glistening glands may aid in attracting insects and other small animals which are caught by the sticky gland-fluid as on a living fly-paper. The struggles of a caught animal stimulate nearby hairs to bend over and to excrete more fluid. The animal becomes finally enmeshed in bent glandular hairs and dies. Digestive juices are then excreted by the glands and these break down the softer portions of the victim's body, the products being absorbed and utilized by the sundew.

The bladderwort (*Utricularia vulgaris*) is an aquatic plant whose vegetative parts are completely submerged. There is little or no structural distinction between branching stem and compound leaf, while roots are absent. On the leaves there are hollow structures known as bladders or utricles which, though small, are visible to the naked eye and are very numerous. Every bladder has a small aperture closed by a flap, which acts as a valve. The inner wall of a bladder has numerous 4-armed hairs which probably excrete digestive substances and also have absorptive powers. The bladder acts as a trap which when set reacts to a mechanical stimulus in such a manner as to suck in any nearby small object such as a water animal or Alga. Such entrapped organisms are not poisoned or by other means done violently to death, but as they cannot escape they eventually die in the trap and their bodies provide food materials for the bladderwort.

Parasitic Plants.

A large proportion of the species of Bacteria and Fungi are parasitic and show adaptations to this mode of life. Since appreciation of their adaptations would involve considerable detailed description of structure and life-history outside the scope of this work, we will take a few instructive examples from flowering plants.

The "partial parasites" are plants which are attached to a host but which have green chlorophyll and so are able to manufacture a certain amount of food for themselves. These include the mistletoe (*Viscum album*) and members of the Rhinanthoideae tribe of the figwort family (Scrophulariaceae). Amongst the partial parasites there is some range in dependence on a host. Thus the mistletoe has no roots and is completely dependent on the tree into which it sends its suckers (haustoria) for the whole of its water and raw food materials ; cow-wheats (*Melampyrum* spp.) do not develop fully unless they become attached to a host ; yellow-rattles (*Alectorolophus* spp. ; also named *Rhinanthus* spp.) and eyebrights (*Euphrasia* spp.) can complete their development, though usually as rather stunted specimens, without a host. None of the green partial parasites shows very much reduction in its aerial vegetative parts but the root systems are poorly developed in the Rhinanthoideae and there is experimental evidence that the absorption of water and raw food materials from the soil is insufficient for full development of the shoots. Normally, in the wild, the partial parasites become attached by side roots to roots of grasses and other plants into which they send absorptive structures.

Peculiarities in structure in correlation with complete parasitism is much more marked. Thus in the broomrapes (*Orobanche* spp.) chlorophyll is absent, the leaves are reduced to scales, and germination of seeds will only occur in contact with living potential host plant roots. It is especially in germination and attachment to the host that adaptive structures and behaviour are evident. A filamentous seedling is first produced, one end of which coming into contact with a host root sends a sucker down into its wood. Absorptive tissues then develop in the seedling and a very close connection is engendered between the different tissues of parasite and host. The other part of the seedling grows into a small tuber from which secondary suckers and a flowering stem arise, or sometimes several stems arise from the tuber. Flowering stems may also develop on roots.

The dodders (*Cuscuta* spp.) show even more marked adaptations to parasitism. Dodder seeds usually germinate in late spring to produce fine yellow threads. The tip of a seedling is raised above the soil surface and grows forward, at the expense of the hinder parts of the seedling, in a spiral manner, so that it appears to creep forward. Seedlings can exist in this manner, at least in some species, for several weeks. If a seedling meets a suitable host it twines round it closely in a series of loose coils alternating with several tight ones. From the latter, papillae, which have developed on the side in contact with the host stem, grow out as suckers which penetrate into the host tissues, spreading out fanwise and linking the conducting tissues of host and parasite. The mature dodder plant is entirely dependent on its host for water and food. It has no root system in contact with the soil, there are only minute scale leaves, and chlorophyll is absent ; yet it produces abundance of flowers and seeds. The dodder belongs to the bindweed family (Convolvulaceae), in which the twining habit is very common. It is interesting to speculate how this may have led to the evolution of the very complete parasitism of the dodders.

PLATE VII

John Markham

BOG ASPHODEL, *Narthecium ossifragum* (Liliaceae), growing through Sphagnum on the edge of a Highland loch Rothiemurchus, Inverness-shire. July 1947

PLATE VIII

Robert Adam

Pinguicula vulgaris (Lentibulariaceae), Ben Ledi, Perthshire, July 1946

The most important adaptations to parasitism are probably the physiological ones and relatively little is yet known about them. There must be special ferment (enzyme) excretions which bring about the entry of suckers into host tissues and also special physiological arrangements for absorption of water and food materials from the conducting strands of the host. It is very probable that some at least of these physiological adaptations are very specific. Certainly, in parasites, partial or complete, there are " physiological races," that is variants with little or no morphological differentiation one from another but restricted to certain kinds of host plants. In the mistletoe three such races have been distinguished, one limited to broad-leaved trees, one to pines and larch, and one to firs. In *Euphrasia* (the eyebright genus) and other genera of the Rhinanthoideae a considerable number of microspecies and microvarieties have been named, but the possible correlation between them and limited circles of host plants has not been sufficiently studied. In *Orobanche* there is also a field for research in the relationship between parasite and specific hosts. Thus, *Orobanche hederae* is restricted to ivy and *O. major* to knapweeds, mainly to *Centaurea scabiosa*. On the other hand, *O. minor* is recorded from nearly a hundred host species, though within this taxonomic species there are some " races," and probably more than yet realized, limited to or showing preference for one particular kind of host. In *Cuscuta* there is the same diversity of behaviour. *Cuscuta trifolii* is usually limited to clovers but *C. epithymum* occurs on a wide range of vascular plants. The whole subject of "physiological races" in flowering plant parasites needs experimental investigation.

One of the most interesting series of structural adaptations concerns the way in which plants pass through the unfavourable season of their annual cycle. In our British climate this is the winter, because of low temperatures. A classification has been made in which the major classes are distinguished according to the position of the stem growing points. It is a fascinating winter study to examine all the plants in a locality (and garden plants too) and by means of many examples to extend the following outline scheme of life-forms into details :

1. *Phanerophytes* : with buds destined to form new shoots projecting freely into the air. This group includes trees and many shrubs. The oak, beech, box, and hawthorn may be taken as examples (Pl. XIXb, p. 234).

2. *Chamaephytes* : with the surviving buds situated close to the ground, whether the branches bearing the buds be woody, as in shepherd's

thyme (*Thymus serpyllum*, in the broad sense), crowberry (*Empetrum nigrum*), and heather (*Calluna vulgaris*), or herbaceous, as in creeping jenny (*Lysimachia nummularia*), stonecrop (*Sedum acre*), common mouse-eared chickweed (*Cerastium vulgatum*), and ground-ivy (*Glechoma hederacea*) (Pl. XX, p. 235).

3. *Hemicryptophytes*: with the surviving buds actually in the soil surface protected by the soil itself and by the dry dead portions of the plant. This is, in our flora, the largest class and the most diverse. It includes daisy (*Bellis perennis*), plantains (*Plantago media* and other species), bladder campion (*Silene cucubalus*), and many grasses and sedges (Pls. XVIII, p. 227, and XXI, p. 242).

4. *Geophytes*: with surviving buds or shoot-apices on shoots some distance below the surface of the ground as in herb paris (*Paris quadrifolia*), coltsfoot (*Tussilago farfara*), fritillary (*Fritillaria meleagris*), winter aconite (*Eranthis hyemalis*), lords-and-ladies (*Arum maculatum*), Solomon's seal (*Polygonatum multiflorum*), autumn crocus (*Colchicum autumnale*), and daffodil (*Narcissus pseudo-narcissus*) (Pl. XXIIa, p. 243).

5. *Helophytes and Hydrophytes*: with the surviving buds protected by lying in mud and water-saturated soil or at the bottom of water respectively. Examples are: sweet flag (*Acorus calamus*), arrow-head (*Sagittaria sagittifolia*), frogbit (*Hydrocharis morsus-ranae*), and pond-weeds (*Potamogeton* spp.).

6. *Therophytes*: with a bud dormant in a seed over the unfavourable period, but completing their life-cycle from seed to seed in one favourable season. This class includes the annuals, such as *Silene conica*, groundsel (*Senecio vulgaris*) (Pl. XIXa, p. 234), goosegrass (*Galium aparine*), ivy-leaved speedwell (*Veronica hederifolia*), and many others.

As already indicated, the British flora is, from the standpoint of life-forms, a hemicryptophytic flora, in that the hemicryptophytes constitute by far the largest class. This is in agreement with the temperate nature of our climate, with rain at all seasons, relatively mild winters, and no deep long-lying snowfalls.

Adaptation in the Flowering Phase of Flowering Plants

A large proportion of our common plants, especially of the herbs, are known best by their flowers. These attract us by form, grouping, colour, and odour, but their adaptation is not to human aestheticism. Flowers have two functions : to produce and protect the reproductive structures and to bring about the union of the male and female sex cells or gametes. After flowering, part or parts of the flower grow into the fruit, but it is most convenient to consider fruits and seeds under another heading. There are many structural adaptations connected with problems of pollination. Pollen-grains contain the male sex cells and have to be deposited on the stigmatic surface of an ovary before the grains can send out pollen-tubes which grow down to the ovary, one to the mouth of one ovule, one to the mouth of another, and so on according to the number of ovules present in an ovary (from 1 to 10,000 or more in British plants). Down the pollen-tubes move the male sex cells, eventually reaching and fusing with the female sex cells (one in each ovule) to bring about fertilization. Pollination can be studied with the naked eye or a hand-lens, fertilization can only be investigated by laboratory methods and a compound microscope. Fundamentally, pollination is merely a preliminary to fertilization, but a necessary preliminary.

In some plants cross-pollination (followed by cross-fertilization) is obligatory and in many it is the rule or is common. By cross-pollination is meant that the pollen comes from one individual plant and reaches the stigma in a flower of another of the same species. In contrast, self-pollination, which is by no means rare, is the deposition of pollen produced by one plant on stigmata of flowers of the same plant. There are many variants of cross- and self-pollination and many interesting problems linked up with the survival values of the result of cross- and self-fertilization. For the moment we may take observed facts and consider them from the well-established conclusions that pollination is (with few exceptions) necessary to the production of seeds and that plants can generally be classified into one or other of two classes : habitually cross-pollinated and habitually self-pollinated. We can reasonably assume that for some species cross-pollination is advantageous, and for some we know it is obligatory since they will not set seed with their own pollen.

Ovules are fixed and stationary inside ovaries. Pollen-grains become free from the anthers (with few exceptions) either as separate dust-like particles or adhering together in groups, but are not self-motile. The simplest method of pollination is the falling, by gravitation, of pollen-grains from an anther on to a stigma or the touching of stigma and anther through approximation in growth or alteration of relative position at maturity. Most often, however, some external agent other than, or in addition to, gravitation or approximation is involved and it is to this that conspicuous adaptations can be observed. Wind and insects are the most important cross-pollinating agents in our British flora and it is instructive to compare adaptations to these two agencies. In general terms (to some of which there are a few exceptions) we may draw the following contrasts :

Wind-pollination	*Insect-pollination*
Flowers often in long or loose inflorescences (as catkins or panicles)	Flowers often solitary or in stiff inflorescences
Perianth absent, or reduced and inconspicuous	Perianth usually conspicuous
Lack of colour, odour, and nectar in the flowers	Flowers coloured, often with odour and nectar
Stamens on long slender filaments	Stamens often on short filaments
Great abundance of pollen produced in numerous or long anthers	Pollen often in smaller quantity
Pollen-grains light, dry, and smooth	Pollen-grains heavier, commonly adhesive, through possession of spines or other protuberances, or by production of viscid substances
Stigmata large, well exserted, often feathery	Stigmata smaller, and often not well exserted

Let us take the poplars and willows as a concrete example. These belong to the same family (Salicaceae) and both have male and female flowers grouped in catkins, the male and female on different trees or shrubs. The poplars are wind-pollinated and have long dangling catkins, especially the male ones, easily moved by a slight breeze : there are numerous (30 to 50) stamens in each male flower ; there is no odour and no nectar is produced; the stigmata are large and branched. The willows are insect-pollinated and have more or less stiff catkins which are often erect ; there are few (2 to 5) stamens in each male flower ; the flowers often have odour and there are well-formed nectaries ; the stigmata are simple. Many of the willows flower before the leaves are formed and this makes the catkins conspicuous, though there is no showy perianth to individual flowers.

The abundant pollen produced by wind-pollinated plants compensates for the great waste which this method of pollination involves, since " the wind bloweth where it listeth." On the other hand it must not be forgotten that pollen is food for certain insects and some insect-pollinated flowers are visited for pollen and not for nectar, good examples being the poppies (*Papaver* spp.) and rock-roses (*Helianthemum* spp.) which have large numbers of stamens per flower. Pollen often " contaminates " honey stored by insects and the honey of the hive-bee owes its vitamin content to the pollen it contains. A large number of wind-pollinated flowers have only one or a very few ovules in an ovary, while many insect-pollinated flowers have numerous (sometimes very large numbers) of ovules in an ovary. There are exceptions to this generalization, but it holds sufficiently often to suggest that the correlation is based on the greater certainty of insect pollination. A comparison of grasses and orchids is interesting in this connection. Our British grasses are all wind-pollinated and have large anthers which at maturity rock to and fro on long slender exserted filaments. The feathery stigmata are also exserted on top of an ovary with only a single ovule. Orchids by their bright colours, conspicuous shapes, and nectar production are adapted to insect-pollination. Actually many orchids are very precisely adapted to pollination by only one kind (or by very few kinds) of insect. Their pollen-grains are united together into usually club-shaped masses (known as pollinia) and a single ovary has a large number (10,000 or more in, e.g., *Orchis mascula*) of ovules. One or two pollinia deposited on a receptive stigmatic surface can pollinate this enormous number of ovules. It appears a gamble, all eggs put in one basket or rather all pollen-grains in (most often) two pollinia. Yet the gamble usually comes off and the relative rarity of many of our orchids is due to other causes than low seed-production. We may include one particularly striking example of adaptation here. Some bee-orchids (*Ophrys* spp.) have a lip the size, shape, and colour of the abdomen of the female of a definite species of humble-bee. Such flowers are visited by males of the respective species of humble bee which, in a number of careful observations, have been found to simulate copulation with the orchid lip. Pollinia become attached to the head of the humble bee and pollination results when they are deposited on the stigmatic surfaces of another flower.

Many flowers are visited by insects for their nectar. In flowers the nectaries may be associated with the receptacle (as in mallows and members of the daisy family), with the calyx (as in cultivated shrubs of the genus *Coronilla*), with the petals (as in buttercups), with the stamens (as in the pink family), or with the carpels (as in marsh marigold and the onion genus). Some flowers have flower parts changed entirely into nectaries. Thus in monkshood (*Aconitum anglicum*) the brightly coloured part of the flower is the calyx. If one removes the back sepal (the " hood ") one finds underneath it two stalked nectaries and these are modified petals. It may be noted that only humble-bees can reach the nectar and in doing so they attach pollen to their bodies or deposit it from their bodies on to the stigmata. Such is the dependence of the monkshoods on humble-bees that the genus does not extend its range beyond that of humble-bees.

In the sages (*Salvia pratensis* can be taken as an example) the two-lipped flower is attractive to bees. Nectar is excreted at the base of the corolla-tube and a bee to reach it must alight on the lower lip, push its head into the mouth of the corolla, and its proboscis down the tube. In doing this it comes into contact with a lever-mechanism if the flower be young (at the " male " stage) and the half-anthers of two stamens are brought down on to the back of the insect's head where pollen-grains are deposited. If the insect now visits a somewhat older flower (at the " female " stage) curved elongation of the style will have brought the stigmatic surfaces to such a position that the bee brushes its head against them just where the stamens of the younger flower had deposited pollen.

There are two risks to the efficiency of adaptation for cross-pollination in some plants whose flowers have very specialized adaptations to pollinating visits of only one kind of insect. The species is obviously dependent for pollination on the presence in sufficient numbers of its one insect pollinator. This presence may be prevented by factors having nothing to do with the plant (bad weather, disease, pests, etc.) and the plant species may have its range consequently limited. Again, biting insects (or other pests) may get at the nectar (or pollen) by some short-cut which does not bring about pollination. Plants with spurred nectaries not infrequently have these bitten through and small insects often crawl into well-concealed nectaries. There are some neat problems in the balancing of the advantages and disadvantages of specialization in adaptation. Extreme specialization may have a temporary advantage, but in the long run it may prove to be over-specialization and over-adaptation so that the species loses any initial advantage and declines in numbers even to extinction.

Some British plants (as clovers and mulleins) are more or less sterile with their own pollen. Inherent causes of self-sterility will be discussed in Chapter 12. Besides such internal control of self-fertilization, there are also external controls of self-pollination, especially such as render it non-habitual. Firstly, there is a separation of stamens and ovaries on different plants (as in willows and poplars). Secondly, there is shedding of pollen and maturing of stigmata at different times even when both occur in the same flower. Thus, in the parsley family (Umbelliferae) and pink family (Caryophyllaceae) the pollen is shed in any one flower before the stigmata of that flower become receptive. Contrasting with this, in autumn crocus (*Colchicum autumnale*) and horse-chestnut (*Aesculus hippocastanum*) the stigmata are receptive first and wither before the pollen-grains of the same flower are shed. Thirdly, there is the mechanism known as heterostylism. A typical example is the primrose, in which all the flowers on any one plant are either "pin-eyed" or "thrum-eyed." In the former the stigma is at about the level of the corolla mouth and the 5 stamens (they are nearly sessile anthers) are about half-way down the tube. In the latter the anthers are at the corolla mouth and the stigma about half-way down. An insect pushing its proboscis down the corolla tube will collect pollen at such a level on its proboscis as to deposit it on the stigma of a flower of the opposite variant (see p. 225).

Cleistogamy. We have seen that in the majority of British plants there are various devices which ensure at least a high degree of cross-pollination. On the other hand there are some species, including very common ones, that are habitually self-pollinated. Some species make the best of both kinds of pollination in the same individual plants by producing two kinds of flowers. An extreme example is seen in the violets, in both the sweet violet (*Viola odorata*) and dog-violets (*V. riviniana*, *V. reichenbachiana*, *V. hirta*, and *V. canina*). The conspicuous coloured flowers of these species are in every way adapted to cross-pollination by insect visitors and also have devices to prevent self-pollination. Seed-setting, however, from these flowers is irregular and sometimes meagre. If one examines a plant, soon after its conspicuous flowering is completed, by delving low down amongst the leaves, one finds numerous short-stalked "flower-buds." Actually, these are not merely young flowers in an early stage of development but flowers which have reduced outer parts, remain small, never open, and most important of all, are self-pollinated. In these cleistogamic flowers, as they are called, the pollen-grains germinate *in situ*, tubes growing out from the anthers to the stigma and thence down into the ovary where the ovules are fertilized. Seed is set in abundance. In an earlier chapter (p. 75) reference was made to photoperiodism. There is experimental evidence that in some species of violets (including the dog-violets named above) the length of daylight is one of the most important of the external factors controlling the production of cleistogamic flowers which in nature only appear, in general, when the light period exceeds 17 hours, that is in summer.

Other plants besides violets produce cleistogamic flowers. Some foreign species reproduce, by seeds, only from cleistogamic flowers, while, on the other hand, there are quite a number of species in which self-pollination occurs in otherwise ordinary flower-buds when extremely adverse conditions (drought, too much water, low temperature, high temperature, shade, etc.) prevent the full development and normal opening of the flowers. Though there is no hard and fast line between such examples of "false cleistogamy" and such a well-marked example as that of the violets, it is important to note that in the latter the flowers are very definitely specialized in structure, especially by reduction of parts, for functioning cleistogamically. Amongst British plants usually producing cleistogamic flowers, we may mention the annual rush (*Juncus bufonius*), the chickweed (*Stellaria media*), the wood-sorrel (*Oxalis acetosella*), and a deadnettle (*Lamium amplexicaule*). Cleistogamy has been recorded for over 50 species of grasses (including many that are non-British). An interesting British species of grass is *Sieglingia decumbens*, which is common on the acid soils of heaths and moors. This plant produces panicles of flowers which are most often cleistogamic. In addition, at the base of the fertile shoots and enveloped within the old leaf-sheaths, there occurs a sessile spikelet which is never exposed and produces a single grain cleistogamically.

Cross-pollination has no significance apart from cross-fertilization. Since there are very varied, and sometimes very elaborate devices, to ensure cross-pollination one would suppose that cross-fertilization has considerable selective advantages. These can only be summed up in the mixing of germ-plasms resulting in variation among offspring.

Unless there be variation there cannot be adaptation and, though variation may initially be due to changes in the genes, crossing and segregation can make new combinations which give variants on which natural selection can act. No variation means no further adaptation to other or changing environments ; too much variation means instability and therefore, for a species as a whole, a failure of adaptation to either a relatively stable environment or to one changing slowly, as well as the possible loss of a vigorous and well-balanced constitution already achieved. There can be too little or too much variation, as well as too high a degree of uncertainty in effectiveness of cross-pollination mechanisms. This probably explains the long-range selective value of combined self- and cross-pollination mechanisms. A very large number of plant species can and do set seed both with their own pollen and with pollen from other individuals. A surprisingly large number of British plants have structural and behaviour devices for giving a possibility or probability of cross-pollination in early stages of flower opening (anthesis) followed later by a reasonable certainty of self-pollination if cross-pollination fails. Complete dioecism, cleistogamy, and self-sterility are exceptional in British flowering plants, so far as observations and experiments have been recorded. When they occur there must be means both of ensuring viable pollination and a balance between other chances of too little and too much variation. Again, there is very frequently an overlap of or imperfection in devices restricting pollination, and this overlap or imperfection maintains some sort of balance between crossing and selfing. Thus, in flowers highly adapted to insect visitors, cross-pollination will be the rule especially when weather at flowering is suitable, but self-pollination may be brought about mechanically as by high winds knocking anthers against or jerking pollen on to stigmata. Insects sometimes visit a flower again and again and may just as easily self-pollinate as cross-pollinate it if anthers and stigmata mature together. In many flowers with either anthers or stigmata maturing first, there is an overlap period of longer or shorter duration either for one or for neighbouring flowers. Even when a plant is " self-sterile " it frequently happens that the self-sterility (whatever the causes) is not absolute, an occasional selfing is successful, and often a self-fertile variety is found within the usually self-sterile species (as in *Trifolium* and *Centaurea*). Such markedly dioecious species (with male and female flowers on different plants) as the large stinging nettle (*Urtica dioica*) and dog's mercury (*Mercurialis perennis*) are occasionally monoecious (with male and female flowers on the same individual plant) with the possibility of selfings. The wall pellitory (*Parietaria diffusa*) shows a mixture of overlapping devices for pollination. The flowers are hermaphrodite, male, or female, in the same individual. The anthers have an explosive mechanism in which the four filaments are at maturity curved inwards and downwards under tension and kept in position between the perianth segments and the ovary or ovary vestige. They expand and spring out suddenly, the anthers burst open by longitudinal splits and the pollen is scattered into the air. Pollination of "own" brushlike stigmata in hermaphrodite flowers is prevented by the stigmata maturing well before the anthers and shrivelling before the pollen is shed, but there is plenty of overlap in maturity of anthers and stigmata between different flowers of the same plant.

Fruits and Seeds

In the seed-bearing plants, pollination, if followed by fertilization, is the preliminary to the production of seeds. The important reproductive part of a seed is the living embryo but the production of this is not the only result of fertilization. Most often, the stimulus of fertilization is needed for the production of the exterior trappings of a seed and for the production of a fruit. Both of these often serve a dual purpose—to protect the embryo during development, and for a period when the seed is mature but before it germinates and to aid in the dispersal of the seeds. It is advantageous to survival of the species for its seeds to be widely scattered. This obviates their germination in masses near together with the seedlings having no room to develop and at the same time extends their range as much as possible. Dispersal mechanisms are often far from perfect and it does sometimes happen that seedlings develop in close proximity, with results well worth studying. Structurally fruits and seeds are different organs, but sometimes a fruit, or a part of a fruit, and sometimes a seed is the dispersed unit, and for our present purpose the mechanism and efficiency of dispersal are more important than the structural relationships of the dispersal unit itself. Like pollen-grains, fruits and seeds are not self-motile and depend on external agents for dispersal for any great distance from the parent plant. Only a few British plants have so-called self-dispersal mechanisms. Wind, animals, and water are the principal dispersal agents for fruits and seeds and adaptations enabling these to act are very numerous and varied. Sometimes there is a combination of methods. It is of some importance to note that the structures which can be considered as adaptations are produced or matured for the most part only after fertilization, after the sex cells have functioned, that their characteristics are genetically controlled, and that, therefore, they are not characters acquired by direct action of the environment.

The field naturalist will find the study of fruits and seeds, their dispersal, the germination of seeds, and establishment of the seedlings, a fascinating study, the more so since there is still much to be discovered in this subject for our British plants. The study can be continuous throughout the year, and can be carried on by a combination of observation with outdoor and indoor experiments. Actual observations and results are required, not *a priori* guesses from structure or analogy. There is such a wealth of examples and suggestions that only a few can be briefly outlined here.

DISPERSAL

Dispersal by the parent plant. Some members of the pea family, as the vetches (*Vicia* spp.) or wild peas (*Lathyrus* spp.), have pods which separate into two halves each of which twists on drying to shoot out the seeds for a foot or so at most. The gorse (*Ulex europaeus*) has pods which open, in dry sunny weather, very suddenly and shoot out the seeds with a noticeable popping noise. The valves of the three-valved fruit of species of violets (*Viola* spp.) are boat-shaped and on drying the margins of a valve come together, pressing out the seeds with some force. The fruits of species of crane's-bill (*Geranium* spp.) eject a single seed violently from every one of five valves. The mechanism is interesting and can be studied in so common a plant as herb robert (*Geranium robertianum*). The long " bill " of the fruit is structurally the persistent and enlarged stylar column. At maturity the lower two-thirds above each one-seeded compartment splits away from the compact central portion. The seeds become detached, but each remains in a carpellary pocket attached by two threads to the corresponding stylar strip. The stylar strip acts as a spring and when a certain degree of tension is attained by the drying-out process it suddenly curls up and breaks away from the central column, with such force that the partial fruit with a seed at the bottom is shot for a distance of about seven yards, the two threads being ripped off upwards from the strip and carried away with the carpellary pocket and seed. It has been shown that the threads aid further dispersal by ants and the garden snail. It is extremely interesting to note how, in the British species of *Geranium*, there is a play upon the central theme of seed ejection which can be tabulated as follows :

Carpels not detached, not wrinkled or reticulate. Seeds thrown and reticulate or pitted : *Geranium silvaticum* " *pratense* " *rotundifolium* " *columbinum* " *dissectum*	Carpellary pockets thrown, wrinkled or reticulate. Seeds smooth : *Geranium robertianum* " *purpureum* " *lucidum* " *phaeum* " *molle*
Carpellary pockets thrown, not wrinkled or reticulate. Seeds smooth : *Geranium pusillum* " *pyrenaicum*	Carpellary pockets thrown, not wrinkled or reticulate. Seeds very finely pitted : *Geranium sanguineum*

Dispersal by wind. Adaptations for wind dispersal are frequently recorded under the headings : tumble-weeds ; lightweight (dust) fruit, seeds, and spores ; winged fruits ; winged seeds ; plumed fruits ; plumed seeds ; woolly fruits ; and woolly seeds. On the whole, winged fruits and seeds are probably dispersed the least widely of these groups, though seeds of Scots pine have been recorded as travelling 880 yards. The light powder or dust fruits and seeds are most easily carried by wind and the evidence strongly favours the conclusion that this group is, on the average, by far the most widely dispersed. It is interesting to note that a very similar mechanism can be developed by fruits and by seeds for wind dispersal (see table on p. 171). More striking examples occur in foreign floras. Sometimes bracts, perianth-parts, or other structures than those developed from carpels or seed-coats make the wings or plumes : as in cotton-grass (*Eriophorum* spp.), limes (*Tilia* spp.), and hornbeam (*Carpinus betulus*).

True tumble-weeds, in which the whole plant or a good-sized portion of it, is blown along by the wind as a regular method of dispersal, occur

	Fruits	*Seeds*
Lightweight	Wormwood (*Balanophora*, an exotic, is the best example)	Orchids
Winged	Birches	Stocks
Plumed	Clematis	Willowherbs
Woolly	Pasque flower	Willows, Poplars

mainly in steppe and more or less desert areas. Dead infructescences with some unshed seeds approximate to this method of dispersal and sometimes occur in grasses, in sea bladder-campion (*Silene maritima*), and clovers.

Very lightweight seeds are usually produced in large numbers. The wind is a wasteful agent, it disperses widely, drastically, and even dramatically, but " where it listeth." Large numbers of seeds are an insurance against heavy or total loss while allowing for the advantages of very wide dispersal. Lightness is often due not merely to small size but also to the development of a loose dry seed-coat.

By exposing suitably prepared glass slides at high altitudes on mountains or in aeroplanes the presence of viable spores of Bacteria, Fungi, mosses, and ferns high in the atmosphere has been proved. That even the lightest seeds can be transported long distances by high air currents remains a reasonable deduction still needing confirmation by direct observation.

The common heather (*Calluna vulgaris*) provides an example in which seed dispersal is, in part, controlled by another organ than the fruit proper, namely the calyx. The sepals in the heather are larger than the corolla lobes but are coloured like them. They persist after seed-setting and four of the five bend over the ripening capsule (dry opening fruit), enclosing it and the other shrivelled remains of the flower. The fifth remains more or less spreading, thus leaving a gap in the upper part of the calycine container. The horizontally placed ripe capsule splits right down the septa but leaves a central column of placentas (places of ovule or seed attachment) and style. The seeds, usually 20 to 30 per capsule, are fairly small and light, but, in so low-growing a plant, would tend to fall direct to the ground were it not for the persistent enclosing calyx, which holds them inside the old flowers till sufficient wind arises to shake them out through the opening in the upper part of the calycine box. Such a wind is strong enough to bring about at least moderately wide dispersal of the seeds, which mainly occurs in the autumn.

Dispersal by animals. Many plants have structural devices by which their seeds are carried longer or shorter distances by animals. Almost all groups of animals aid in the dispersal of seeds but birds and mammals are the most important. It is usual to distinguish between internal and external carriage. The edible portions of fleshy fruits—like berries, pomes, and drupes—enclose a seed or seeds protected by a hard covering which may be either an inner layer of the fruit or the seed-coat. It is remarkable how very similar adaptational ends are attained by extremely diverse structural developments. Even within the narrow limits of the British flora, we find the edible portion of the fruit developed from the middle ovary wall as in wild cherries or blackberries, from the inner ovary wall and placentas (places of attachment of ovules or seeds) as in berries like the nightshades (*Solanum* spp.), from receptacles as in dog-rose or wild strawberry, from receptacle and ovary wall as in pomes like the crab-apple, or from an aril (a growth round the seed) in the yew, while if well-known cultivated plants be added we have the juicy perianth of the mulberry and the fleshy infructescence of the fig. The functioning of these dispersal mechanisms follows one general principle : the flesh is digested while the seeds, protected by a hard covering, are indigestible and are passed through the alimentary system and out with the excreta, unharmed and sometimes in a better condition for germination. Again we see that differences in the kinds and behaviour of effective external agents, combined with diversity of variation, direct evolution through natural selection to functionally similar ends by diverse routes.

External carriage is also made possible, and even probable, in many species by a great variety of structural devices—hooks, hairs, bristles, spines, development of mucilage, and so on. Here is yet another example of similar ends attained by the special adaptation of different structures acting in essentially the same way. Again, the range of structures could be extended and more striking examples given if non-British plants were included. We will take hooks which catch in fur or feathers:

Hooked bracts in burdocks (*Arctium* spp.)

Hooks on fruiting calyx in forget-me-nots (*Myosotis arvensis* and other species) and bur-marigolds (*Bidens* spp.)
Hooked styles in buttercups (*Ranunculus* spp.) and wood avens (*Geum urbanum*)
Hooks on fruit wall (inferior ovaries, the outer part bearing the hooks usually regarded as receptacular) in enchanter's nightshade (*Circaea lutetiana*), wood sanicle (*Sanicula europaea*), and goosegrass (*Galium aparine*)
Hooks on fruit wall (superior ovaries, hooks from true ovary wall) in small medick (*Medicago minima*)
Hooks on seed-coat. These are very rare and no British example is known

The maximum efficiency of seed dispersal by animals is limited by the distances from the parent plant covered by any carrier. With the possible exception of the distance covered by birds, this is usually not great in Britain. It is, however, important to realize that wild mammals were formerly more varied in kind, and the farther-roaming ones much more numerous, before man destroyed or modified so much of our wild animal life and indirectly in this way, and more directly in many other ways, modified our flora and vegetation.

There is a good deal of observational and some experimental evidence that birds disperse seeds very effectively within their territories or feeding-areas both by internal and external carriage. An interesting modification of the former, by the way, is the gorging on fleshy fruits, flight to some distance, and then ejection of the pulped mass through the mouth—the bird apparently overeats and is sick. There seems rather a tendency for certain birds so to behave when a plentiful crop of ripe yew berries is available, as can be observed in the neighbourhood of Box Hill. Ducks and game-birds, spending much time on mud or soil, have been found to have seeds attached to their feet sometimes in adhering mud or soil. Water birds are probably important agents in the stocking of isolated ponds with marsh and aquatic plants. Carnivorous birds and such as grind down hard fruits and seeds with grit in their stomachs are obviously of no direct importance in internal dispersal. The possibility of very long-distance dispersal by birds on migration is a controversial subject on which it is very difficult to obtain satisfactory direct evidence. Statements, based on examination of migrating birds killed at lighthouses, have been made to the effect that birds travel empty of food and with well-cleaned feet and feathers.

If this be substantiated, birds cannot be considered as agents dispersing seeds over long distances of sea, as to oceanic islands.

Dispersal by water. A number of our coastal seed-bearing plants are dispersed at least partly by the waves and currents of the sea. The seeds and vegetative parts of salt-marsh plants, as we have already seen, are not harmed by immersion in salt water. We have, of course, nothing in our flora comparable with the coco de mer or double coconut of the Seychelles (*Lodoicea maldivica*), whose fruit is one of the largest known, and coconut (*Cocos nucifera*) whose fruits with their fibrous and woody covering are dispersed by ocean currents and by man. It is interesting to recall that seeds of West Indian plants are not infrequently cast up on our western coasts, whither they have travelled with the Gulf Stream, and sometimes they are viable, but the species cannot establish themselves in our climate. Examples of British species disseminated along our coast, by floating seeds or drifting branches, are sea-kale (*Crambe maritima*), sea rocket (*Cakile maritima*), sea-beet (*Beta maritima*), and glassworts or samphires (*Salicornia* spp.). A great many freshwater aquatic and marsh plants have devices for water dispersal. It has been estimated, on the basis of experiments, that of fruits and seeds of British species, 60 per cent sink immediately or within a week, 25 per cent float for a month or more, and 15 per cent float for over 6 months, and to this last group belong those which can float through the winter on our ponds and rivers. Again we find various structures involved in special adaptations, which for water-transport take the form of loose or spongy coverings retentive of air. Such include the utricles (bracteoles) of sedges, the glumes (bracts) of marsh grasses, the sepals of water-docks, corky rings surrounding the nutlets in gipsy-wort (*Lycopus europaeus*), fruit-walls in some water buttercups and water-plantain (*Alisma plantago*), and corky or air-containing tissues in the seed-coat as in yellow flag (*Iris pseudacorus*) and bog-bean (*Menyanthes trifoliata*), persistent swollen parts of the seed (raphe and chalaza) in marsh marigold (*Caltha palustris*), or very light seed-leaves (cotyledons) in sea purslane (*Honckenya peploides*). For both freshwater and sea dispersal, buoyancy and retention of vitality with submersion are essential, and the seeds, or seeds plus fruits, of a large proportion of salt-marsh, freshwater marsh, and aquatic plants combine these features. In freshwater plants in particular, dispersal of vegetative parts is common and germinating seedlings are also transported without risk of desiccation.

Anne Jackson

BLACKTHORN, *Prunus spinosa* (Rosaceae). Witherslack, Westmorland, near Grange-over-Sands. April 1942

PLATE X

Eric Hosking

Alternative or additional methods. Through not infrequent accidents, seeds may be dispersed by a method which by-passes the method for which there is structural adaptation. Thus, the floating seeds of water-plants may become stranded on mud and adhere, with some mud, to the feet of ducks or other water birds and so get transported from one pond or river to another. An interesting double adaptation is shown by the common gorse. As already noted, the seeds are shot out by the sudden opening of the pods and the twisting of the valves. As they lie on the ground they are often dragged away by ants attracted by the brightly coloured caruncle (protuberance near the attachment point of the seed) filled with oily food-material. Some of the seeds are dropped by the ants on the way to their nests and dispersal to a distance from the parent plant is thus effected.

Numbers of Seeds Produced and Their Viability

The number of seeds produced, per flower or per plant, varies enormously from species to species and from individual to individual of the same species. Most quantitative work has so far been done on annual or biennial herbs, and there are few or no sound estimates for trees or shrubs or even totals for the life of perennial herbs. A few figures will indicate the kind of range in seed-production for single plants of:

Lepturus filiformis (a grass)	90 seeds
Limonium binervosum (a sea-lavender)	400 to 1,000 seeds
Rumex crispus var. *trigranulatus* (a dock)	4,000 to 25,000 seeds
Glaucium luteum (yellow-horned poppy)	24,000 seeds
Linaria vulgaris (toadflax)	29,000 seeds
Beta maritima (sea-beet)	130,000 seeds
Verbascum thapsus (mullein)	700,000 seeds
Digitalis purpurea (foxglove)	750,000 seeds

In dealing with numbers of seeds there are, for those annual or biennial species producing more than one seed per fruit (simple or compound), two variables : the number of seeds per fruit and the number of fruits per plant. The mean of the products of the mean values of these two variables, plus or minus their probable errors, gives what is known as the *average seed output*. However, the number of seeds produced is not necessarily an index of viability. Again, figures will illustrate this:

In *Mercurialis perennis* (dog's mercury)	5–15	per cent of the seeds are viable
,, *Mercurialis annua* (annual mercury)	70	,, ,, ,, ,, ,, ,, ,,
,, *Alnus glutinosa* (alder)	25	,, ,, ,, ,, ,, ,, ,,
,, *Betula pendula* (birch)	25	,, ,, ,, ,, ,, ,, ,,
,, *Fraxinus excelsior* (ash)	90	,, ,, ,, ,, ,, ,, ,,
,, *Verbascum thapsus* (mullein)	88	,, ,, ,, ,, ,, ,, ,,

The product of the average seed output and the fraction represented by the average percentage germination gives the *average reproductive capacity*.

A number of important points, relevant to selection, have emerged from recent research on seeds and seedlings. Large seeds are advantageous in species otherwise suited for growing in dense plant communities. In such, enough food supply must be provided for the seedling till it grows to a height where sufficient light is available for photosynthesis (food manufacture by green colouring matter in light). Correspondingly, small seeds are often found in plants normally growing in open habitats. In general, the larger the supply of food in the seed the more advanced the phase of succession the species normally occupies. A wide variety of evidence shows that a high reproductive capacity is a positive asset, not only insuring numerical replacement of deceased individuals but providing for the successful invasion of additional territory. Adverse conditions, unless extreme, cause a diminution in the number of seeds produced by a plant but do not affect the quality of those matured. Competition affects seed output greatly, and in considering reproductive capacity it is essential to carry out studies on individuals growing under natural conditions. Actual and potential achievement are two different phenomena of different significance in biological investigations and particularly in natural selection. Little is known regarding the natural percentage survival of viable seeds up to germination. There is probably a great variation in seed mortality.

Germination

Germination is the beginning and early stage of growth of the embryo (young plant) in a seed. Moisture and a certain range of temperature are external essentials for the germination of all seeds, and for its successful continuation oxygen is also required. The seeds of most of our British plants germinate in the spring when there is usually sufficient moisture available and rising temperatures provide the necessary warmth. Apart from its convenience as a dispersal mechanism, the seed is admirably adapted to increase the chances of survival of plants growing on land and therefore subjected to the risks of cold and drought at certain seasons of the year. Dormant seeds, that is those alive but not germinating, contain much less water than most plant parts. Largely because of their relatively dry condition, they can withstand very great extremes of heat and cold and lack of moisture. Oxygen is necessary for normal respiration, which is essentially a combustion (oxidation) of food materials with the release of energy and is characteristic of living cells in general. Water is absorbed by seeds as a preliminary to true germination and this absorption is a reversible process—the seeds can be dried out again without injury. Once germination proper has commenced, by growth of the young root of the embryo, there is no going back to a state of dormancy. If moisture, heat, and oxygen are present, energy is obtained by normal respiration and utilized in growth. Any long-continued absence or great reduction of any one of these essentials results in death. A characteristic of some germinating seeds, which obviously has survival-value under certain conditions, is their power of continuing growth for a period of at least some days in the absence of oxygen. This anaerobic respiration, as it is termed, allows germinating seeds to grow for a period in water-logged or compacted soil, perhaps till their growing tips can reach a supply of air.

The survival-value of very early spring germination followed by rapid growth to maturity is well seen in members of the ephemeral flora of some dry places with such plants as mouse-eared chickweeds (*Cerastium* spp.), rue-leaved saxifrage (*Saxifraga tridactylites*), hairy bittercress (*Cardamine hirsuta*), and two forget-me-nots (*Myosotis collina* and *M. versicolor*). Summer droughts, even of short duration, have very marked effects on wall-tops and sand-dunes where these plants often grow. By germinating very early and setting seed in a few weeks they are " adapted " to droughts by avoiding them (see, too, p. 148).

A very wide range of seed behaviour is found in connection with the time that must or can elapse between the setting of the seed by the parent plant and its germination. Some seeds have to germinate almost immediately or they lose their vitality. The seeds of some species of willow (*Salix*) can only germinate within one week of being shed. The seeds of poplar (*Populus*) must also germinate very soon after maturity or they will die. Acorns lose the power of germination within a few months. Thus, in the British flora there are examples of short-lived seeds in both small-seeded and large-seeded groups. There are considerable risks of undue loss of functioning seeds by these plants though the risk is probably on the whole balanced by the large number of seeds produced by willows and poplars and by the very low reproductive rate essential to maintain an oakwood or even to extend its range. In other species, perhaps in a large number, germination can occur immediately the seeds are shed by the parent plant, or even before they become free, but may be delayed for a longer or shorter period. In rose-bay willowherb (*Epilobium angustifolium*) seeds have been found germinating in the capsules after the parent plant had been blown down, but 20 per cent germination was obtained of collected seeds six months after they had been gathered. Seedlings of the dandelion (*Taraxacum officinale*, in the broad sense) have been raised eight days after removal of the fruits from the parent plant, though some of the seeds will retain their vitality for two to three years when kept dry. Seeds which retain their vitality over long periods, germinating when conditions are most suitable, give the species producing them particular advantages of survival. The species of the British flora whose seeds can retain their power of germination for the greatest number of years is, so far as reliable records have been traced, the kidney-vetch (*Anthyllis vulneraria*), whose hard seeds have germinated after 90 years.

Some species have seeds which naturally always remain dormant for longer or shorter periods after shedding. There are a number of causes of such dormancy, with similar end results, obtained by different mechanisms. The seeds may have rudimentary or immature embryos when they are shed, and germination can only occur after the embryos have become fully developed, as in lesser celandine (*Ranunculus ficaria*) and herb-paris (*Paris quadrifolia*). Many plants of the pea family, and some other plants, have seed-coats which prevent the intake of water in the so-called " hard-seeds." Scratching the surface (scarifying, as it is called) permits water to enter and germination to start. Sometimes the embryo is incapable of rupturing the tough seed-coat, through which, however, water can permeate. Thus, in the water-plantain (*Alisma plantago*) the seed-coat has, in nature, to decay or receive

Box, *Buxus sempervirens* (Buxaceae). Box Hill, Surrey. October 1946

Brian Perkins

F. Ballard

a. Bladder Campion, *Silene cucubalus* (Caryophyllaceae). Leatherhead, Surrey. June 1946

John Markham

b. Sea Bladder Campion, *Silene maritima* (Caryophyllaceae)
Varieties with and without anthocyanin
On shingle bed, R. Spey, near Aviemore, Inverness-shire. June 1946

damage before the embryo can pierce it, though artificially released embryos grow at once. Retardation of gaseous exchange preventing the growth of the embryo is stated to be the cause of dormancy in cockle-burs (*Xanthium* spp.) (of which some occur as aliens in this country), some other composites, and a wild oat (*Avena fatua*). The seeds of some hawthorns (*Crataegus* spp.) and some junipers (*Juniperus* spp.) need what is termed " after-ripening " before they will germinate. In one American hawthorn it has been shown that the principal change in after-ripening is an increase in the acidity of the cell-sap in certain portions of the embryo. It is probable that this acidity produces conditions which favour the absorption of water and the formation and activity of ferments (enzymes) which stimulate growth.

A classification of germination behaviour in seeds has been suggested as follows :

(*a*) simultaneous germination, in which the seeds all germinate within a comparatively short period of one another, as in stork's-bill (*Erodium cicutarium*), small-flowered buttercup (*Ranunculus parviflorus*), and ivy-leaved speedwell (*Veronica hederifolia*) ;

(*b*) continuous germination, where germination within a batch of seeds continues over a long time, without obvious breaks as long as external conditions are favourable, as in corn gromwell (*Lithospermum arvense*) ;

(*c*) discontinuous germination, in which germination occurs at markedly separated intervals, as in gean (*Prunus avium*), flowering rush (*Butomus umbellatus*), and a rockrose (*Helianthemum breweri*).

It would appear that, other conditions being equal, there is more risk to the species as a whole in having simultaneous germination than in having either continuous or discontinuous germination. The survival-value of discontinuous germination is increased in that the spreading-out in time which characterizes it gives the opportunity of filling space, as this becomes vacant from deaths of earlier established plants of the same generation.

There are a number of external conditions which must be present for the germination of the seeds of certain species but to which the seeds of the majority of species are indifferent. Light is one of these. Light-sensitive seeds, i.e. seeds which need light for germination, include those of mistletoe (*Viscum album*), willow-herbs (*Epilobium* spp.), many speedwells (*Veronica* spp.), and purple loosestrife (*Lythrum salicaria*). Light-hard seeds, which have their germination prevented or retarded by light, include those of species of onion (*Allium* spp.) and some commonly cultivated garden flowers (*Nigella sativa*, *Phlox drummondii*, and *Nemophila insignis*, amongst others). The selective value, if any, of such behaviour may differ from one species to another. The seeds of some species require freezing before they will germinate. The bog-bean (*Menyanthes trifoliata*) and moschatel (*Adoxa moschatellina*) are examples. Seeds of various root-parasites, as those of broomrapes (*Orobanche* spp.) and toothwort (*Lathraea squamaria*), are said to germinate only in the neighbourhood of roots of suitable hosts. This would give some advantage to the maintenance of dormancy for a long period of dispersal, with increased chances of reaching a host. Orchid seeds, in nature, can only germinate successfully when infected from the soil by a fungus which combines with the embryonic roots to form mycorrhizas.

Seedlings and Seedling Establishment

Seed production, embryo viability, and germination are, for any generation, preliminaries to seedling establishment. The seedling is one of the most vulnerable phases in the life-history of a plant. Reproductive capacity gives an incomplete picture without some knowledge of its translation into reproductive actuality in natural survival of viable seeds, their germination, and the establishment of the seedling. Seedlings are subject to high mortality rates. It remains doubtful how far this is selective of advantageous variations and how far indiscriminate. Much is certainly indiscriminate. Given the external conditions already mentioned, seeds will germinate whether conditions suitable for establishment of seedlings be present or not. The Biblical parable of the sower is ecologically sound. It is to be noted that seedlings are less complex structurally than are adult plants. There are fewer morphological characters on which natural selection can act. On the other hand, some physiological variations, in particular rate of growth and disease resistance, may be of considerable importance in this connection. The subject requires much more investigation and, in particular, close observation of the actual causes of death in seedlings in the wild. We know that certain mutations are heavily selected against in seedlings, notably those resulting in chlorophyll deficiencies, which range from albinism (complete failure to develop green plastids) to various degrees of imperfect development of the chloroplastids or of their contained chlorophyll. Albinos naturally die as soon as the food reserve in the seed is used up and chlorophyll-deficient seedlings are quickly at a disadvantage in competition with normal seedlings. Seedlings of parasites and saprophytes form an exception to these statements.

There is sometimes an obvious connection between the size and shape of the fruit, the size, shape, number, placentation (method and place of attachment), and arrangement of the seeds, the size, shape, and packing of the embryo, and the details of germination and structure of the seedling. It does not, however, seem reasonable to say that any one of these features determines the other. Given genetically fixed characters at any stage of ontogeny (development of the individual) there must be conformation at other stages within a limited range of variation, and there will be selection, usually drastic, against variants outside this range.

The food supply necessary to the seedling before it becomes self-supporting is an instructive example of a need met by different means and the special adaptation of different structures, as the following examples show :

Food reserve in pre-fertilization or prothallial endosperm	in Coniferae
" " " perisperm	in Caryophyllaceae
" " " post-fertilization endosperm	in Gramineae
" " " cotyledons	in Leguminosae
" " " hypocotyl	in *Bertholletia* (Brazil Nut)

With relatively few exceptions the seed-leaves (cotyledons) are undivided and much simpler in form than the foliage leaves of the same adult plant. This is largely an adaptation to packing in the seed. It is also an aid to establishment of seedlings, since it facilitates penetration of covering soil without damage and enables the seed-leaves to act as organs for absorbing food supplies in germinating seeds with endosperm, or to expand speedily for food manufacture. There are also adaptations in the protection of the young growing point of seedlings. In monocotyledons the rule is for the one cotyledon to penetrate the soil while the stem growing point is protected by its lateral position, only just above the cotyledon base. The growing point of dicotyledons is between and more or less enfolded by the two cotyledons. Seedling tips are often bent before they emerge from the soil.

General Remarks on Adaptation and Natural Selection

The examples of adaptations given above are only a small fraction of those that could have been quoted. They have been deliberately chosen to show the wide range of the subject. All kinds of plants show adaptation to their natural environments; all kinds of organs show adaptations with reference to their particular functions; every phase of the individual's life-history involves adaptation if there is to be survival to the next phase. Yet every variation is not an adaptation. Two facts, or groups of facts, have to be remembered in this connection: the individual plant, most particularly in its adult life, is a highly complex unit, and evolution, even within the flowering plants, has proceeded through millions of years, and natural selection has acted more or less vigorously throughout this time under general conditions not greatly, if at all, dissimilar from those now existing on the earth. Combined consideration of these two facts leads to the conclusion that any genetic variation (mutation) is much more likely to result in the appearance of a disadvantageous than an advantageous character. Many mutations of this sort have been recorded. Let us consider one example. The common bladder campion (*Silene cucubalus*) (Plate 38a, p. 179) and the sea bladder campion (*S. maritima*) (Pl. 38b, p. 179) belong to the subfamily Silenoideae of the family Caryophyllaceae. The distinguishing structural character of the subfamily is that the sepals are united into a tube (which is blown-out and bladdery in the species under discussion), only short teeth remaining free. The petals and stamens arise below the ovary, the former having long and usually well-defined claws and the latter long and slender filaments (stalks). There is an evident and necessary correlation of characters: the tubular calyx protects the developing petals, stamens, and gynoecium (female part of the flower), but, for these to function, parts of them must project beyond the calyx mouth. Mutations affecting any of the floral parts may upset the structural balance of the whole floral adaptation—which fundamentally is an adaptation to ensure pollination. Three such mutations which have been found in the campions may be mentioned. One, recorded several times in the wild and studied experimentally in both *S. maritima* and *S. cucubalus*, involves a failure of the petal claws to develop to a sufficient length by the time the other flower organs are mature. The consequence is that the " poor " petals more or less fail to emerge from the normal calyx and their blades are very poorly developed (they may even be entire, small, and greenish). A second mutation, found only two or three times in *S. maritima*, results in a much-elongated cylindrical calyx-tube. The inner floral parts fail to emerge from this and, in one extreme plant, there was complete failure to set seed. A third mutation, only found once in *S. cucubalus*, was most instructive. In this plant the calyx was split down to the base more or less completely into free-spreading sepals. The petals were normal but the flower was female with no functional stamens, a condition not uncommon in otherwise quite normal plants. The petals sprawled outwards with nothing to support them—the whole floral arrangement and pollination mechanism was upset. It is interesting, though not immediately relevant, to note that the first and last mentioned mutations could be considered as partial reversions to an earlier ancestral condition. If the two were combined in one individual resulting in separate spreading sepals and clawless or very short-clawed petals (and shortened filaments) a condition similar to that in the other subfamily (Alsinoideae) of the Caryophyllaceae would result. At least, it seems a fair deduction from the observations and experiments made with these plants, that any one of the three mutations mentioned would, by itself, be soon eliminated because the floral structure makes pollination extremely difficult and thus reduces seed-setting and reproductive capacity.

The observant field naturalist will come across many individual plants with aberrant structures—extreme examples may be termed " sports "—and simple experiment may show they have a genetic basis. Very frequently it can be shown that they involve an upset in the normal routine life-history of the plant, to such an extent that survival either of the individual or of its offspring is jeopardized. Other mutations, however, are not so immediately dangerous to continued existence of plants possessing them and selection only acts against them in the long run. Observations on populations containing such mutations are much needed over a period of many years and would yield invaluable data if fully recorded. Finally, one meets with characters, genetically fixed, that appear to be unselected. It is always rash to be quite sure that a character has no advantage under any environmental conditions, but it does seem probable that characters neither advantageous nor disadvantageous to the species in its normal mode of life are not infrequent. We will take two examples from the campions mentioned in the last paragraph. In *Silene maritima* (Pl. 38b, p. 179) the leaves and stems are always glabrous, but in every good-sized population of *S. cucubalus* (Pl. 38a, p. 179) examined in Britain glabrous and densely hairy individuals have been found. These breed true on selfing and on crossing the hybrids have an intermediate development of hairs and themselves segregate on selfing or back-crossing. There is nothing in the distribution or behaviour of the variants to indicate that there is any adaptational value in having glabrous or hairy stems and leaves. In both *S. cucubalus* and *S. maritima* there is a pair of contrasting characters of the seed-coat, since this may be either " tubercled " (with small tubercle-like projections) or " armadillo " (with flat angular plate-like markings). " Armadillo " is always recessive to " tubercled," in each species and in interspecific hybrids, whichever way crosses are made. No exception has been found to this rule in a very large number of experiments. Yet populations of *S. maritima* occur in the British Isles in which a majority, often a very large majority or even all, of the plants have armadillo seeds while other populations have only tubercle-seeded plants or these in a majority. In *S. cucubalus*, in this country, most populations have a large proportion of the plants with tubercled seeds (often about 90 per cent). So far no explanation on the basis of natural selection has been found for the recorded facts.

There are several useful terms, and the concepts they designate, which can be introduced here with advantage. Plants of similar appearance, or, more technically, having the same characters, are said to belong to one " phenotype." Plants with the same gene constitution are said to belong to one " genotype." To consider a phenotype we look at a plant, to consider a genotype we have to know the breeding behaviour of the plant.

The term "ecotype" is now being used rather freely for certain groups of plants within a species. The concept behind the term is both valid and useful, but it requires some attention to understand its exact significance. Varieties within a species are differentiated one from another by characters having a genetic basis, or, in other words, varieties belong to different genotypes. Ecotypes are intraspecific groups distinguishable one from another not only by having different genes but by having characters such as make them adapted to different habitat conditions. Plants within one ecotype may belong to more than one genotype so long as the different genotypes are not limited in ecological range to distinct habitats. An ecotype is thus the product of natural selection, acting on the mixture of genotypes within a taxonomic species in such a manner as to relegate one set of genotypes to one habitat, another to a different habitat, and so on. Good examples of ecotypes are *Solanum dulcamara* var. *marinum* and *Cytisus scoparius* var. *prostratus*, both limited to the sea coast.

It is sometimes found that a species (or other taxonomic group) varies along a definite direction within its geographical or ecological range, the variants, when plotted, showing as a gradient. The term cline has been given to such a gradient. The sea plantain (*Plantago maritima*) shows clines in relation both to climatic differences in different parts of its geographical range and to local habitat differences. The ground pine (*Ajuga chamaepitys*) (Pl. 39, p. 186) shows a large-scale cline in its geographical range from Asia Minor across Europe to south-eastern England. Careful search should be made for clines (especially ecological clines) amongst British plants.

There are other structures which are not adaptational under the conditions in which the species now naturally lives, but which are maintained in the species population (they have not been selected against to the extent of extinction) and are either post-adaptational or pre-adaptational. Vestigial structures were, one can reasonably assume, at one time of value to the species, but owing to changes, either in the living conditions or in the more general structure of the plant, or in both, they became useless. There is a general tendency for organs not used to become reduced in size and in complexity and even to become represented by more or less indeterminate vestiges with loss of function, though in some instances there may be a neatly balanced argument as to whether they have become reduced because they have ceased to function or have ceased to function because they have become reduced. There are also examples of many kinds and of all degrees of change of function. Scale leaves and staminodes (vestiges of stamens) provide examples.

The term pre-adaptation has been used to designate structure or behaviour which predisposes a plant to take advantage of a certain type of environment or adapts it from the outset to particular conditions. Thus the annual or ephemeral habit when already present in a species is often a pre-adaptation to conditions found in early stages of plant succession. Actually, with this rather vague meaning, structure or behaviour in many or all examples of successful invasion, migration, plant succession, and so on, could be described as examples of pre-adaptation. Most of such examples are, however, adaptations of long standing, though changes of environment may have meant a new relationship of the old adaptations to more or less new conditions.

Over-adaptation, in the sense of extreme specialization to a narrow range of conditions, is probably a cause of extinction of species or of paramorphs (groups within species) and, in the course of geological time, of larger groups. Adaptation of pollination mechanism to a single species of insect, or of a parasite to a single host-species, means dependence on the occurrence at the right time and place of one particular other organism, and failure to reproduce if the pollinator or host cease to be available.

PLATE 39

Brian Perkins

GROUND PINE, *Ajuga chamaepitys* (Labiatae). Chipstead valley, Surrey. July 1946

PLATE 40

F. Ballard

BLACKBERRY, *Rubus fruticosus* agg. (Rosaceae), Bladon Heath, near Woodstock, Oxfordshire, August 1944

CHAPTER 12

THE STUDY OF HEREDITY IN BRITISH PLANTS BY GENETICAL AND CYTOLOGICAL METHODS

THE NUMBER of species of the wild British flora that has been investigated genetically by experiment, especially by the use of localized British material, is decidedly smaller than might be anticipated. Some cytological research has been done on a larger proportion of the species, but many of the records require checking and extending as they are based on non-British material not always adequately determined. It seems best for us to take various examples and to deal with some briefly and with others more fully. These examples follow the sequence of the Bentham and Hooker classification. Reference may be made to Appendix F for a brief outline of Mendelian principles and to Appendix E for a summary explanation of cytological terms and concepts.

RANUNCULACEAE

RANUNCULUS FICARIA (lesser celandine). Two variants of this species have been studied in Britain and the results published. The diploid plant has $2n=16$ chromosomes and usually produces a variable proportion (averaging nearly 50 per cent of the carpels) of viable seeds. It has no tubers formed in the axils of the stem leaves. The var. *bulbifera* has $2n=32$ chromosomes and very rarely produces apparently "good" seeds (only 2–3 per cent of the carpels swell and appear to ripen) and germination tests showed that only 18 per cent of these were viable. It multiplies almost entirely by vegetative means—by the root tubers formed at the base of the plant just below ground level and by small root tubercles formed in the axils of the stem leaves. Failure to set seed is immediately due to various forms of degeneration in the embryo-sac or nucellus (the part of the ovule in which the embryo-sac is embedded), not to lack of viable pollen. There is some evidence that the two variants have ranges which are at least unequal. In parts of Middlesex, Surrey, and Buckinghamshire both occur, but var. *bulbifera* is the commoner. In parts of Devon and Somerset only the variant without stem tubercles has been found. It would be very interesting to have records from all parts of Britain on this matter for the species is recorded from all vice-counties. Apomixis (setting of seed without fertilization) occurs in the species. Other variations in *R. ficaria* including variations in sex, petal colour, leaf shape, and leaf markings have been studied experimentally but the results have not yet been published.

RANUNCULUS AURICOMUS (goldilocks). This species of woodlands and half-shaded habitats has two well-marked varieties in this country: the one without petals (apetalous) and the other with five well-developed yellow petals making a large buttercup flower. In the former the leaves and particularly the radical leaves are either not divided at all or are only sparsely dissected or lobed ; the latter has finely dissected leaves.

The sepals of the apetalous variety are more delicate, wider, usually slightly crumpled, yellow on the inside, but still green on the outside. This subpetaloid character of the sepals is interesting since in some other Ranunculaceae, for example *Caltha palustris* (marsh marigold or kingcup), there are no petals and the sepals are fully petaloid (though they have no nectaries). Reciprocal crossings of the two varieties led to some interesting results. The families obtained from the two crosses were different, even in the seedling stage. All the offspring raised when the petaloid variety was the ovule parent and the apetalous variety the pollen parent showed the dissected leaves of the maternal plant, while the offspring obtained when the apetalous variety was the female parent and the petaloid the pollen parent started with undivided leaves and the later ones were only lobed or sparsely divided. The hybrid plants with dissected leaves produced flowers mainly intermediate in character, with 1 to 4 petals (rarely 5). In the F_2 generation a few completely apetalous flowers were observed on one plant. In the cross with the apetalous variety as the female parent, the flowers, like the leaves, exhibited entirely the characters of the female parent, except that a single flower produced one petal. Thus, while fertilization of the petaloid variety with pollen of the apetalous variety yields somewhat intermediate offspring, the reciprocal cross shows purely unilateral (metromorphic) inheritance. A conclusive explanation of the above facts has still to be found. Hybrid populations are rather frequent in the wild, at least in the south and midlands of England, and in them various strange floral abnormalities occur, including transitions between stamens and petals, petals and sepals, and sepals and bracts, suppressed stamens (functionally female flowers), and malformed carpels—one of the latter was found with an extruded ovule. A common abnormality in one such population involved the formation of tubular nectaries in the position of the petals, strongly recalling the normal nectaries in *Helleborus* and *Eranthis*. Chromosome counts have given numbers $n = 8$, 16, and 24 in continental stocks, but no British material appears to have been examined cytologically.

RANUNCULUS ACRIS (common buttercup). As might be expected in so widespread and abundant a species there is considerable variation in many characters. Leaf-blotch has been shown to have a genetic basis, but leaf shapes have apparently not been studied genetically in British material. Petal colours (lemon chrome, ivory yellow, lemon yellow) have been shown to be gene-inherited. The sex of the flowers raises interesting problems on which experiments have thrown some light. Mention has already been made (p. 135) of the occurrence of plants with hermaphrodite, female, male, and neuter flowers. In every large-sized population examined (such as a field of buttercups) hermaphrodite and female plants have been found, though the proportions vary. There are also intermediates between hermaphrodite and female. Neuter plants have only been discovered, in the wild, two or three times, and a male plant only once, though many have been bred by using the pollen of this one on other stocks as female parents. The interpretation of breeding results is complicated by the occurrence of some apomixis and, in some stocks, a degree of self-sterility. The male condition is completely recessive to the hermaphrodite and segregates, with all its peculiarities, in F_2 in a ratio probably representing 3 : 1. Neuter plants are always constant but cannot, of course, be used in breeding. Plants intermediate, and fluctuating, between female and hermaphrodite are of hybrid constitution. The figures obtained, from controlled selfings, suggest the following genetic scheme :

Full hermaphrodite	MM FF
Full female	mm FF
Full male	MM ff
Neuter	mm ff

Intermediate and fluctuating plants, or plants whose use as parents results in segregation, have a mixed constitution, such as MmFf, MMFf, MmFF (all of which were tested). All the plants examined cytologically in this research had $2n = 14$.

RANUNCULUS BULBOSUS (bulbous buttercup). Less research has been done on this species but it is interesting to record that a parallel series of flower colours to those of *R. acris* (lemon chrome, ivory yellow, and lemon yellow) has been shown to have a genetic basis. Female plants occur in the wild.

Papaveraceae

PAPAVER RHOEAS. This common cornfield weed requires further experimental study. Supposed hybrids between it and *P. dubium* have been recorded but experimental proof of their correct status is lacking. Two well-marked variants involve the hairs on the flower-stalks ; in one they are spreading and in the other adpressed. Both variants can usually be found in large populations, but most often, in Britain, the latter are in a minority. Experiments with Danish stocks showed that the characters of adpressed hairs is dominant over spreading hairs, though there was a tendency for the recessive type to segregate in numbers slightly exceeding expectations. The adpressed-haired plants flower on the whole earlier than the ones with spreading hairs. The greater frequency of the recessives in wild populations (in Denmark and other parts of the Continent, as in Britain) has been tentatively explained by its supposed higher vitality under northern conditions. The poppy is probably of Mediterranean origin and preliminary observations suggest that in parts, at least, of the Mediterranean region the adpressed-haired variant (dominant) is much commoner, relative to its allelomorph, than it is in Britain (or Denmark). A race hardier for more north-western conditions may be in course of selection. The genetical experiments need extending to British material. Conditions intermediate between spreading and adpressed hairs occur in the wild and are not explained by the above results.

Flower colour has been studied genetically in poppies and the following scheme has been suggested to explain the colours found in wild and cultivated stocks of *P. rhoeas* (in which the Shirley poppy is included) :

A linkage group with :

B, producing dark filaments and a dark centre.
W, producing a white edge on coloured flowers.
R, producing a slight dilution of flower colour.

John Markham

RAMSONS, *Allium ursinum* (Liliaceae), in hazel coppice near Wells, Somerset
May 1946

John Markham

WHITEBEAM, *Sorbus aria* (Rosaceae). Edge of Dunstable Downs, Hertfordshire October 1945

A second linkage group with :

- P, converting purple flowers into red.
- T, in the absence of B, producing tinged filaments and darkening the flower colour, but lightening the flower colour in the presence of B.
- F, producing large flushed white centres in the absence of B.

An independent factor, C, is recorded in the absence of which the flower ranges from white to variously dilute and striated forms as compared with the full colour in the presence of C.

Other variable characters which await genetical study are shape and depth of division of the leaf, shape of capsule, and number of stigmatic rays. The species would form useful material for an experimental study of the overlapping characters correlated with different genetic constitutions and those found in different environments.

Cruciferae

NASTURTIUM spp. (water-cress). Recent cytological investigation has shown that there are two species of water-cress in Britain. The one (*Nasturtium officinale*) is diploid with 2n = 32 ; the other (*N. uniseriatum* or *N. microphyllum*) is tetraploid with 2n = 64. There are no reliable and easily visible differences in the vegetative parts for distinguishing the two species, but the fruits and seeds are very different. In the diploid, the fruits are shorter and relatively broader than in the tetraploid ; the seeds are in two ranks with large reticulations, about 25 of which are visible on each side. In the tetraploid, the fruits are long and narrow, with the seeds in one rank with very small reticulations, almost pitted in appearance, with about 100 visible on each side. There is fairly good evidence that the stomatal index[1] on each surface of the leaf is different for the two species (in *N. officinale* 17·7 and 15·0 and in *N. uniseriatum* 11·2 and 10·8, for lower and upper leaf surfaces respectively). Hybrids are not uncommon between the two species. They are triploid (2n = 48) and sterile. The distributions of the two species and of the hybrid require further study. The little evidence available suggests that the diploid has, on the whole, a more southern and the tetraploid a more northern range, but there is a considerable

[1] The stomatal index is the ratio of the number of stomata to the total number of epidermal initials, according to the formula $S.I = \frac{S}{E + S} \times 100$, where E = number of epidermal cells and S the number of stomata in a unit area.

overlap. In England both occur and field studies in Oxfordshire failed to show any differences between them in ecological preferences. In the River Glyme and around Blenheim Lake, for example, both occurred in random distribution, together with the hybrid. The cytological evidence indicates that the tetraploid is an allopolyploid and probably arose as a cross between *Nasturtium officinale* and *Cardamine flexuosa*, though attempts to cross these two species have so far failed.

EROPHILA spp. The whitlow grass (the collective species is usually known as *Erophila verna*) is well known to be represented by a large number of kinds which have been variously described as species, subspecies, and varieties. These different kinds (or some of them) are genetically constant, self-pollinated, and self-fertilized, and local populations are usually very uniform in taxonomic characters. In a recent investigation chromosome numbers have been found as follows :

Chromosome number :	7	12	15	16	17	18	20	26	27	29	32	Totals
Denmark	33	—	22	1	1	1	—	6	—	2	—	76
England	—	—	11	—	2	3	1	—	—	—	—	17
Scotland	—	—	—	—	—	—	—	4	2	—	—	6
Holland	1	—	1	—	—	—	—	—	—	—	1	3
Germany	—	3	—	—	—	—	1	—	—	—	—	4
Sweden	—	—	1	—	—	6	—	—	—	—	—	7
Totals :	34	3	35	1	3	20	2	10	2	2	1	113

Hybridization is rare in nature, but it is probable that new kinds arise naturally as a result of hybridization and some new types with new chromosome numbers may be stabilized, following hybridization, by the pronounced self-fertilization of the plants. As a result of close study the collective species *E. verna* has been divided into 4 species, which are distinguishable by a sum-total of characters, including chromosome number, are intra-specifically fertile, but essentially sterile one with another, and have characteristic geographical and local distributions. These species have been named :

Erophila simplex ($n=7$) *E. duplex* ($n=15-20$)
E. semiduplex ($n=12$) *E. quadruplex* ($n=26-32$)

VIOLACEAE

VIOLA spp. The British species of *Viola* are divisible into the violets (sect. *Nominium*) and the pansies (sect. *Melanium*). The classification of the species in each section is difficult because of hybridization, which is often on a considerable scale. Chromosome numbers show a wide range in the genus and different numbers are found in different individuals in some of the accepted species.

We will first consider some recent researches on one of our common dog-violets, *Viola riviniana*. This species, in Great Britain, is generally well marked off from its allies, but is, within the species, heterogeneous. Two subspecies have been accepted as ecotypes: subsp. *minor* and subsp. *nemorosa*, each breeding true for its distinguishing characters. The former is typically an open grassland plant, and the latter a woodland one. The chromosome number recorded for *V. riviniana* by several cytologists is n=20 (or 2n=40, when the material examined was root-tips), but individual plants with 2n=46 and 2n=48 have been recorded. Both subspecies of *V. riviniana* show a considerable amount of variation, often parallel in the two, of a genetical nature but not yet studied in detail by experimental methods. *V. reichenbachiana* (syn. *V. silvestris*) has 2n=20 and hybrids between this species and *V. riviniana* have 2n=30 or 34.

Viola canina, another dog-violet, has been the object of some controversy. A Danish botanist considered it "a cytologically irregular species with *oscillating chromosome number*," since he obtained different numbers in different plants. It has since been suggested that these Danish plants were hybrids. In wild British material from four widely separated localities, the number has been found to be constantly 2n=40.

It is interesting to tabulate the chromosome numbers of violets so far recorded in British material:

2n=20	2n=40	2n=46 or 48
Viola odorata (3 varieties)	*Viola canina*	*Viola riviniana*
V. hirta	*V. riviniana*	
V. reichenbachiana		

Turning to the pansies we find that the cytology and the genetical behaviour, as shown in the results of experimental crossing, are complicated in material of wild origin. It is extremely difficult or even impossible to draw clear-cut lines of taxonomic differentiation and this difficulty is reflected in the cytology. One cytologist, who has studied the group in detail on the Continent, wrote : "The species of the *Melanium* section constitute a complete series of transition as regards intersterility, morphological differences and conjugation of chromosomes in their hybrids." Hybridization is common when different units meet. Anyone who has seen such hybrid swarms cannot but be impressed by the range of character combinations, and the beauty of many of them, or fail to realize the possibilities of the establishment of new units by natural selection from the mixed population. Unlike the violets, the pansies do not normally produce cleistogamous flowers (see p. 165). The marked adaptations to cross-pollination by insects favour hybridization.

A hybrid swarm between *Viola tricolor* (n = 13) and *V. lutea* (n = 24) has been studied in considerable detail. Many combinations of parental characters occurred, though in plants not approximately intermediate there was a tendency to resemble *V. lutea* more than *V. tricolor*. Varying chromosome numbers were found and various irregularities in chromosome behaviour. The reduced n-number ranged from 13 to 27 (mostly it was 23 to 26). There was no doubt that in such a mixed wild population there was a chance for favourable combinations of chromosomes to occur such as would lead to the formation of new phenotypes which might be selected because they possessed advantageous adaptations. On the other hand, malformed abnormalities, in some of which the flowers were "empty shells," were produced. Natural selection would immediately select against these, indeed some of them could not reproduce. It was possible to show, as has also been done in studies on continental pansies, that a large number of characters, some of specific or even higher group value, are determined by genes behaving in a Mendelian manner, though because of chromosome irregularities and gene interactions, regular ratios were not always obtained. In the *V. lutea* × *V. tricolor* crossings such genes included : H, a dominant gene producing hairiness of the leaves : P, a dominant gene for spathulate middle lobe of the stipule ; Y, a dominant gene for long and slender petaloid spur ; N, a dominant gene for dentate sepaline appendages. Other characters are controlled by several interact-

PLATE XIII

Robert Adam

ONE-FLOWERED WINTER-GREEN, *Moneses uniflora* (Pyrolaceae), in old pine wood. Sutherland. June 1933

PLATE XIV

Robert Adam

TWIN FLOWER. *Linnaea borealis* (Caprifoliaceae), forming carpet beneath pine wood. Moray. July 1934

ing genes. The genetical inheritance of petal colours is complex.

It is of interest to add that the cultivated pansies, produced by species-crossing between *Viola lutea* and *V. tricolor*, and between these and *V. cornuta*, have, as a rule, n = 24 (approx.), but the sex cells in one and the same plant may have different chromosome numbers. Some garden varieties show a high degree of sterility. The unstable genetic system in the pansies probably, at least in part, accounts for the rapid deterioration often seen in untended garden populations of pansies. The horticulturally finer plants are unstable or have a greatly reduced reproductive capacity and are replaced, sometimes surprisingly quickly, by reproductively more vigorous and more stable plants, which the horticulturist may despise.

A very fine series of natural hybrids, forming a hybrid swarm, between *Viola variata* and *V. arvensis* occurred in an arable field near Gomshall, Surrey, in May, 1932. The population was a large one of some thousands of plants and there were scores of variants; indeed, apart from the parental types, almost every plant seemed different from every other in one or several characters, and strikingly so in those of flower size, colour, shape, and orientation of the petals. Such a hybrid population would well repay statistical and cyto-genetical analysis.

Caryophyllaceae

SILENE OTITES is a rare species in Britain, where it occurs in four vice-counties in East Anglia (West Suffolk, West and East Norfolk, and Cambridge), but has a wide range in Europe and temperate Asia, at least as an aggregate, composite species. The plants are sub-dioecious, i.e. there are normally male plants and female plants, but occasionally more or less hermaphrodite plants, or intersexes as they are sometimes called, can be found. Crossing experiments have revealed what seems to be a complicated genetical background in the group to which the British *S. otites* belongs. It is rather doubtful if the schemes so far suggested are fully explanatory. However, so far as British material is concerned, *S. otites* is probably heterogametic in the female for sex, that is two kinds of ovules are produced, one kind with a gene-content resulting in female offspring and the other with a gene-content resulting in male offspring, while the pollen-grains are all of one kind in this respect. When hermaphrodites occur, they behave, in breeding, as modified females.

SILENE CUCUBALUS (Pl. 38a, p. 179) and *S. MARITIMA* (Pl. 38b, p. 179). These two bladder campions have been referred to several times already in connection with special problems. Since few if any wild British species have been more fully investigated genetically, it is considered worth while to give a very condensed summary of the results obtained. The summary is limited by two considerations. Firstly, publication of the results of completed experiments was interrupted by the war, and, secondly, a full appreciation of the problems involved, and of their solutions, would necessitate reference to studies of plants from populations in countries outside the British Isles. Only brief references will be made to unpublished results and to foreign materials.

The two species are clearly separable and are maintained as distinct populations in the British Isles. To name them as subspecies or varieties of one species can only confuse the interesting problems connected with their structure, behaviour, and history. Generally they cross reciprocally with each other and give a fertile hybrid. Yet hybrids between them are rare in nature and no large hybrid swarms have been found between them nor has any tendency for merging of populations. In nature they are kept clear-cut and distinct by markedly different ranges of ecological tolerance. Characters studied generally include the following :

Habit. The ascending habit and absence of barren procumbent stems, characters of *S. cucubalus*, are dominant, or nearly so, in some F_1 families, but there is complex segregation in F_2. In other families habit-characters are intermediate in F_1 and give very complex F_2 families. A peculiar strict habit with the leaves nearly erect is occasionally found but is so affected by environmental conditions as to make its genetical investigation difficult.

Indumentum. S. maritima is always glabrous ; *S. cucubalus* may be glabrous or densely hairy on the leaf surfaces and lower parts of the stems. In crosses between densely hairy and glabrous plants the F_1 is intermediate in hairiness and in F_2 an approximation to a ratio of 1 densely hairy : 2 intermediate : 1 glabrous was obtained. This suggests a simple pair of factors with absence of dominance. In addition, however, evidence was obtained of a factor which can reduce the number of hairs to few and another factor (or more than one) which modifies the length of the hairs.

Leaves. Leaf shape and size vary greatly in each species. *S. maritima* has usually decidedly smaller leaves than *S. cucubalus*. A considerable number of stocks breeding true for leaf-characters has been studied in both species. On crossing these *inter se* the F_1 is usually intermediate, while in F_2 segregation occurs, showing that many factor-pairs are involved within and between the two species—probably a dozen or more in all the plants experimented with (including non-British stock-plants). It is interesting to note that local populations occur in the wild in which all the plants have similar leaf shape while in other wild populations there is more or less variation. In *S. cucubalus* a mat surface to the leaves, stems, and calyx is dominant over a shiny surface in F_1 and the segregations obtained in F_2 have given a 3 mat : 1 shining-green ratio, though from selfings in one stock ratios inconsistent with so simple a genetical explanation were obtained.

Inflorescences. The inflorescence in the bladder campions is typically a false dichotomous cyme with a terminal flower to every branch, so that a fully developed inflorescence should have 1, 3, 7, 15, 31, 63, 127 . . . flowers. In *S. maritima* the number of flowers rarely exceeds 7, while in *S. cucubalus* the flowers are usually 31 or more in well-formed inflorescences. Environmental conditions, in the broad sense, very much affect flower production and there is also evidence of cytoplasmic influence as shown by different results in reciprocal crosses. The F_1 usually shows an intermediate number of flowers while in F_2 there is a considerable range in flower number, but when the results are grouped in classes there are modes at 3, 7, 15, and 31.

Anthocyanin development. Anthocyanin may be developed in any, all, or none of the aerial parts : stems, leaves, calyx, corolla, anthers, filaments, stigmata, and young ovules. There are genes for its presence or absence, and for its distribution and intensity. On the other hand, given the presence of the genes necessary for anthocyanin production, environmental factors, such as cold, drought, and disease, can cause an intensification of the colour. Such environmental modifications give a fluctuating series of overlapping phenotypes of different genetical constitution. Plants lacking one or more of the basic genes necessary for anthocyanin production at all are always distinctly characterized by a yellowish-green colour of all their vegetative parts and calyces. They always breed true to absence of anthocyanin on selfing, but the crossing of two such plants may result in anthocyanin in the offspring, or in some of them. A minimum combination of genes is necessary for the production of anthocyanin. Inhibitory factors for anthocyanin production also occur in some plants. It is of interest to note that the full gene complement may be present but unable to express itself as in plants, with only female flowers, carrying genes for anthocyanin in filaments. The genes are there but not the filaments! This condition has been proved by using such plants in extensive breeding. An example of limitation of anthocyanin development to a particular organ, or its absence from that organ, is found in the immature seed-coat. This may be white or purplish. Populations in both species vary greatly in the wild in the percentages of plants with white or purplish immature seeds. Generally speaking white-seeded plants are commoner in *S. cucubalus* and pink-seeded in *S. maritima*. Stocks breeding true for both characters have been isolated. In straight crosses between parents breeding true for white and purplish immature seeds respectively, purplish is dominant to white, but F_2 segregation shows that a number of gene-pairs are involved in colour-production. When only one of these is heterozygous in F_1 a 3 : 1 ratio has been obtained in F_2, but with further heterozygosity less simply interpreted ratios occur. Petals in both *S. maritima* and *S. cucubalus* are normally white. Near Fearnan, on the shores of Loch Tay in Scotland, there is, or was, a population of *S. cucubalus* originally almost entirely of reddish-purple-flowered plants, but with white- and pink-flowered plants, in the course of years, becoming commoner. Breeding showed that selfed heterozygotes segregated in numbers suggesting a 3 coloured to 1 white ratio, and depths of colour were probably, in part at least, due to phenotypic modification of one genotype. Results so far published make it clear that complementary and inhibiting genes are both involved in the presence or absence of anthocyanin. It is probable that one set of genes is basically responsible for anthocyanin development wherever it occurs in the plant. Locally or temporally acting modifiers, however, intensifying, diluting, or inhibiting, prevent, when they are present, uniformity of action by this basic set in all parts developing successionally on a given shoot system of any one individual. Modifying effects are the product of: (1) genes of strictly circumscribed action; (2) genes whose action is less pronounced in later growth-phases of a given shoot; and (3) factors environmental to gene action.

Flower shape. The flowers of *S. maritima* are erect and regular (actinomorphic), those of *S. cucubalus* are nodding and irregular (zygomorphic). In the F_1 the flowers are slightly nodding and less strongly irregular than in *S. cucubalus*. There is segregation in F_2 but the number completely erect and regular is always low (about 1 in 6).

Calyx shape. The " bladdery " calyx is usually larger in *S. maritima* than in *S. cucubalus*, but there is a considerable range in shape and size in both species, especially in the former. In crosses between plants with markedly different calyx shapes, both within and between the species, the F_1 is usually intermediate if the parents are true-breeding for calyx shape. In F_2 there is segregation which varies according to the genetical composition of the grandparents. In a straight cross between *S. maritima* with broadly ellipsoid calyx and *S. cucubalus* with ovoid calyx, the F_1 was intermediate and in F_2 a 1 : 2 : 1 ratio was obtained. A more complicated situation is shown by another experiment within *S. maritima*. A plant with cylindric calyces was crossed with a plant having peculiar long cylindric calyces. The former plant segregated on selfing into plants having ellipsoid, narrow ellipsoid, cylindric, or fluctuating calyces ; while the latter bred true for calyx shape. The F_1 showed segregation into plants with ellipsoid, intermediate, and cylindric calyces. An F_2 family from the selfing of an F_1 plant with ellipsoid calyces gave ellipsoid 27 : intermediate 36 : cylindric 2 : long cylindric 2. Another F_2 family from the selfing of an F_1 plant with cylindric calyces gave ellipsoid 0 : intermediate 39 : cylindric 30 : long cylindric 0. Thus, ellipsoid throws a majority of ellipsoid and intermediate, while cylindric throws only intermediate and cylindric. The true breeding of peculiar long cylindric plants and the very few long cylindric plants appearing in F_2 suggest that long cylindric is in some sense a bottom recessive for the factor group involved.

Petals. The petals in both species have a bilobed lamina on a long narrow claw. In most stocks, the petals of *S. maritima* are larger than those of *S. cucubalus*. Petal characters with a genetic basis include : size, extra lobing, overlapping of segments and of the lamina as a whole, depth of lobing, and degree of development of corona (scale to boss). In *S. maritima* a plant with very small petals (" poor " petals) bred true for this character, but on crossing with two plants breeding true for full petal development gave F_1 families with fully developed petals. Segregation occurred in F_2 in the ratio 3 fully developed : 1 " poor " petal. Another type of small petal occurs in female flowers and is correlated with the factors resulting in femaleness. Multilobing is not infrequent in *S. maritima* in some British populations. Its expression fluctuates considerably from flower to flower, but in some families at least it is recessive to bilobing with segregation occurring in F_2. Overlapping of petals and overlapping of segments can only be studied adequately in well-grown plants with fully developed and completely hermaphrodite flowers and after sunset or on a dull day. Overlapping of petals and overlapping of segments are typically dominant in *S. maritima*. *S. cucubalus* usually has the segments diverging (correlated with the smaller lamina and narrower lobes in this species). Studies on the depth of lobing have involved the use of non-British stock-plants and it will suffice here to say that the character has a genetic basis. The degree of development of the corona raises interesting problems. Typically there are well-developed scales in *S. maritima* and a mere boss in *S. cucubalus*. The F_1 is intermediate and there is segregation usually approximating to a 1 : 2 : 1 ratio in F_2. Nevertheless, in wild populations of *S. maritima* plants occur occasionally with reduced scales though other specific characters are all in accord with the determination of the material as *S. maritima*. These may well be coronal mutations though rarely they may be the last traces left by long-past hybridization with *S. cucubalus*.

Sex. Female plants are not uncommon in both species. They can often be detected by having flowers smaller than are the hermaphrodites. Male flowers have not been found in any British material of either species. A complication in scoring is that some plants produce both female and hermaphrodite flowers, while the number of polliniferous stamens fluctuates in such plants between 0 and 10 per flower. There is some evidence that cytoplasmic factors may be involved ; at least, in many families raised under full control the maternal influence is very marked. The problem of " femaleness " or " male-sterility " in these species of *Silene* awaits a full solution. Its interest is increased by the fact that in some other species of the genus dioecism (male flowers and female flowers on different plants) is the rule.

Capsules. In both species there are several fruit shapes. In interspecific crosses the F_1 shows an intermediate condition between the squat, broadly ovoid or obloid capsule of *S. maritima* with its wide mouth and six reflexed teeth and the ovoid-ellipsoid capsule typical of *S. cucubalus* with its narrow mouth and six erect teeth. In F_2 there is often a range in capsule characters from those of *S. maritima* to those of *S. cucubalus*. However, with very careful scoring the ratio has been found to work out as 1 *maritima* type : 2 various intermediate types : 1 *cucubalus* type. Crosses between variants within the species show overlapping between capsule characters determined by genetical factors and those due to environmental factors—particularly poor nutrition.

Seeds. Two kinds of seeds are clearly distinguishable in each species: "armadillo," with flat plates making up the seed-coat, and "tubercled," with small tubercles covering the seed. Always, in interspecific and intraspecific crosses between armadillo and tubercle types, and in reciprocal crosses, armadillo is recessive to tubercled. Armadillo always breeds true. Tubercled either breeds true or segregates in a 3 : 1 ratio (given large enough numbers and no other complications). There are subsidiary factors which modify the degree of expression of the armadillo or tubercled character respectively.

The above account outlines only some of the results obtained in twenty years' intensive and extensive study of a pair of British species. It is particularly interesting to note that so many of the characters studied genetically, and particularly those considered to have specific value by taxonomists, do not show dominance in the heterozygote; in other words, the F_1, and particularly the interspecific F_1, has an intermediate character. This is markedly true for habit, leaf shape and size, number of flowers, regularity or irregularity of flowers, corona development, and capsule characters. This "failure of dominance" is a common characteristic in hybrids between species of plants, and further supports the view that *S. maritima* and *S. cucubalus* are best considered as distinct species. It also fits in with the observed fact that in nature hybrids between the two species are rare and do not form hybrid swarm populations on any large scale, so that there is little opportunity for selection to act in favour of dominance so far as characters distinguishing the species are concerned. The most striking exception, of complete dominance of tubercled over armadillo seeds, is not correlated with interspecific differences since both types occur in each species. Another interesting fact is that the distinctions in habit between the species have a (complex) genetical basis and are of marked survival-value in the very different habitat ranges of the two species. Here is a specific character genetically inherited and markedly adaptive. One other matter must be interpolated here. Some twenty characters are more or less distinctive of the two species, and using combinations of these there is not the slightest difficulty in determining to which species a given British wild plant belongs, apart from the rare hybrids (and these can similarly be detected by a knowledge of the structure and behaviour of hybrids made artificially).

Nevertheless, one does find, on examining character by character a large number of plants in wild population, individuals which show a single character belonging to the other species. These are probably most often mutations and give some indication of one step in species formation. One example is capsule shape. Thus, in a large population of *S. cucubalus* on the Hog's Back, Guildford, Surrey, one or two plants had capsules which in shape and reflexing of the teeth agreed with capsules of typical *S. maritima*.

MELANDRIUM spp. The two British species of *Melandrium*, *Melandrium album* (the white campion) and *M. dioicum* (or *M. rubrum*, the red campion), are now widely diffused throughout the country, though the former is rarer in the north and the latter much less frequent where the land is highly cultivated and there are few woodlands or brushwoods. It is very possible that the white campion is not native but was introduced, perhaps in Neolithic times, with early cultivated crops, while the red campion is probably a true native. However this may be, when populations of the two meet, hybrid plants or hybrid swarms are common and are readily detected by pink coloration in the petals (there are also other differences between the species and other characters for the determination of hybrids besides petal colour). The production of colour is dominant to its absence, but there is a complication in that anthoxanthin accompanies and acts as a co-pigment to the anthocyanin, causing a blueing of the reddish-purple colour of *M. dioicum*, and the anthoxanthin is present to a greater extent in *M. album* than in *M. dioicum*.

These campions are of considerable interest in that much research has been done on them with regard to the cytogenetics of sex. Both species are normally dioecious, and it is worthy of note that seeds giving rise to male plants germinate, on the average, more quickly than do seeds giving rise to female plants. In the male plant there is a special XY pair and in the female a definite XX pair of chromosomes, in addition to 22 ordinary chromosomes. Sex chromosomes in flowering plants were first reported in these plants. It has been demonstrated that in *Melandrium* sex is determined by the strength of sex genes in the X and Y chromosomes together with some genes in the other chromosomes (autosomes). Further, it has been shown that there are linked genes in the X and Y chromosomes. Various chemical differences have been found between male and female plants in *M. dioicum*.

John Markham

MALLOW, *Malva silvestris* (Malvaceae), showing a fasciated stem 7 inches wide. Broxbourne, Hertfordshire. August 1938

PLATE XVI

G. Atkinson

Female (pistillate) plants are higher in dry weight, in the rosette and blooming stages, and are higher in pH, fresh weight, and oxidase activity in the blooming stage only. Female buds are higher in total sugars, nitrogen, and polysaccharides. Male (staminate) plants are higher in ash, phosphorus, nitrogen, and total sugars in rosette and blooming stages. Male buds are higher in dry weight, ash, and oxidase activity. Recently polyploids have been obtained, in *M. album*, by heat-treatment and by treatment with the drug colchicine, by several investigators. Such treatment results in doubling of the chromosome sets, but mere doubling does not affect the sex expression. If A symbolizes one complete set of 11 ordinary chromosomes (autosomes), X the chromosome with female-producing genes, and Y the chromosome with male-producing genes, the diploid males (2A+XY) become tetraploid males (4A+XXYY) and diploid females (2A+XX) become tetraploid females (4A+XXXX) after treatment with colchicine. On crossing XXXX females and XXYY males the F_1 population consisted of 3XXXX : 87XXXY : 2XXYY, owing to the production of an excess of XY gametes by XXYY males. The XXXY plants constituted a new type with intermediate sex-chromosome balance. They were predominantly male, but about 10 per cent bore a few hermaphrodite flowers. By crossing XXXX females and XXXY males a fertile dioecious race of tetraploids was established. This result is important since it shows that the possession of sex chromosomes is not necessarily a bar to the establishment of fertile polyploids.

LEGUMINOSAE

TRIFOLIUM PRATENSE (red clover). This is a very variable species of considerable economic importance as a hay plant. Three groups of variants have been recognized: early-flowering, late-flowering, and wild. All three now occur in Britain, but it is interesting to note that the cultivated early-flowering red clovers have a general European distribution south of 50° N. latitude while north of 60° N. the late-flowering types are the more prevalent. Red clovers are normally cross-pollinated by bumble bees, and in genetical experiments it has been found best to utilize these natural pollinating agents after they have been given a thorough bath in water to remove or destroy any viable pollen attached to them. This process is made possible by red clover being almost completely self-sterile so that it is not necessary for the tiny florets to be de-anthered, a process which, in any event, is almost impossible without unduly damaging the reproductive parts. The chromosome number in all red clovers examined cytologically has been found to be $2n = 14$, except that tetraploids, with 28 chromosomes in their body cells, have been produced artificially. Hybrid vigour is shown by crosses between variants as compared with in-bred plants.

Amongst the numerous characters proved to have a genetical basis in red clovers the following are included. Central leaflet spot is dominant to its absence. Restriction factors give red stems and green stipules, and *vice versa*. Interaction of factors produces the various flower colours : pure white, yellowish-white, purple-red, pink, blue, and bluish-pink. The usual purple-red colour is known to be due to the interaction of 14 or more dominant genes. At least four factors interact to produce the range of seed-coat colours. Various factors are necessary for normal habit development. Some thirty different types of " dwarfs " and " smalls " have been recorded. A number of recessive genes influencing the growth of specific organs have also been identified. These include 13 factors affecting leaf-growth, one factor governing stem development, and 6 factors responsible for flower development. Resistance to mildew has a genetic basis.

Of particular interest is the very common occurrence of plants with deficient chlorophyll production. 152 recessive chlorophyll-deficient genotypes have been found, mostly determined by different recessive factors inherited on a simple Mendelian basis. Of these, 57 are completely lethal during the early seedling stages—24 being completely or almost completely devoid of chlorophyll, and the other 33 ranging in colour from yellow tinged with green to light green ; while 95 types survived to the adult stage. In addition to these, 8 normal green lethal mutants have been found in red clover. Most of these types, which are intensely dark green, perish in the very young seedling stage, possibly as a result of excessive chlorophyll production.

Another phenomenon which has been investigated in considerable detail in red clover is self-sterility. Growth of pollen tubes is retarded in stylar tissues carrying the same gene as the pollen. This is known as incompatibility. The initial rate of growth of the incompatible and compatible pollen tubes is practically the same, but after traversing about half the length of the style, the growth of the pollen tubes carrying a gene like one in the stylar cells is suddenly retarded and its subsequent growth is exceedingly slow. Very rarely indeed can the ovule be reached (there is only one in each flower) before it disintegrates. There is a whole series of these incompatibility-genes which are conveniently referred to as S_1, S_2, S_3, and so on. They occupy the same position (locus) in the chromosomes, of which there are 7 pairs. A pollen-tube nucleus will have one, and stylar cells have two, of these allelomorphic genes. Thus, a cross (or selfing) :

Female *Male*
$S_1S_2 \times SS_{12}$ will be infertile.
$S_1S_2 \times S_2S_3$ will allow only the pollen with S_3 to function.
$S_1S_2 \times S_3S_4$ will allow any of the pollen grains to function.

There is a very large number of the S-gene allelomorphs in red clover. Thus within two varieties 41 and 37 have been found respectively. On the other hand, there are very rare plants carrying a gene giving complete self-compatibility. This gene, known as S_f, is one of the same series as the other S genes to which it is dominant.

With 7 chromosomes as the haploid number one would expect 7 linkage groups. Five of these have been found : one with 19 genes, one with 6, and three each with 4.

TRIFOLIUM REPENS (white clover). This is not known to be so variable as red clover, possibly because it has not been studied so intensively by genetical methods, and possibly because it propagates so readily by vegetative means through its prostrate easily rooting stems. It may be mentioned that two kinds of plants occur with regard to the presence or absence of a cyanogenetic glucoside (a chemical compound yielding glucose and prussic acid) whose production is genetically controlled by a simple pair of genes. The production of the glucoside is completely, or almost completely, dominant to failure to produce it. In addition there are probably modifying factors which act as intensifiers or diluters of the quantity of glucoside produced. Another character of unusual interest has been shown to have a genetic basis. A plant was found which showed severe mottling of its leaf blades similar to the symptoms caused by virus. This was crossed with an unrelated normal plant. Families were carried on to F_2 and F_3 generations and back-crosses were also made. The results were explained by assuming two independent dominant factors, both of which must be present for the development of the mottling. Here we have an example of heritable factors inducing a character similar to the response induced by certain disease agents. There may be significance in the suggestion, which some investigators have made, that viruses and genes are related. A single factor-difference genetically distinguishes pink- from white-flowered plants, pink being dominant. The pink-flowered variety (var. *erubescens*) is the common variant in the Scilly Isles. There are a large number of allelomorphs for self-incompatibility and independent factors which cause occasional self-fertility (" pseudo-self-fertility ").

ANTHYLLIS VULNERARIA (the kidney vetch) is a variable species, though less so in Britain than on the Continent. It seeds well both on selfing and crossing with other plants of the same or distinct varieties. In nature both homozygous and heterozygous plants occur and, where distinct varieties meet, hybrid swarms are produced between them. There are a number of genes concerned with producing a wide range of colours in the petals which, especially in heterozygous plants, are not necessarily all the same colour in a single flower, since the genes may express themselves differently in standard, wings, and keel. In particular, the var. *coccinea*, typically a variety of coastal areas in the west of Britain, when crossed with var. *lutea* gives a wide range of colour variations, many of which segregate on selfing for two or more generations, but true-breeding new colour variants have been extracted.

Mention should be made of a plant, termed " Amaranth Purple " because it was not matched with any named variant, found at Par Harbour, Cornwall, as an alien. The introduction of this plant resulted in a hybrid swarm consisting mainly of phenotypes totally distinct from those occurring where " Amaranth Purple " was absent. This is an interesting example, carefully studied in the field and in the experimental ground of the influence of a plant introduced by chance modifying a previously established population.

LOTUS CORNICULATUS (bird's-foot trefoil). Like many other kinds of plants *Lotus corniculatus* contains a glucoside which on decomposition through the action of an enzyme or dilute acids yields, among other substances, hydrogen cyanide (HCN) or prussic acid as it is commonly called. As leaves and stems die the enzyme acts on the glucoside and HCN is liberated. It is, of course, in any but very small amounts,

Brian Perkins

WATER AVENS, *Geum rivale* (Rosaceae). Wood Ditton, Cambridgeshire. April 1946

F. Ballard

Dog Rose, *Rosa stylosa* var. *systyla* (Rosaceae). Chessington, Surrey. July 1946

poisonous to animals and poisoning is not infrequently recorded in stock fed on cyanogenetic plants, as those containing the HCN-producing glucoside are called. The HCN is easily detected by change from yellow to red of filter paper soaked in a solution of picrate. In populations of *L. corniculatus* in Britain, plants with and others without the glucoside occur. Cyanogenesis (production of HCN) or the presence of the glucoside is determined by a dominant gene. The species, however, is a tetraploid in the sense that it has four sets of chromosomes, $2n = 24$ being the basic number (*x*-number) for the genus. The inheritance of the dominant gene responsible for cyanogenesis is, therefore, tetrasomic since with 4 chromosomes containing it or its allelomorph a 35 : 1 instead of a 3 : 1 ratio is to be expected in F_2 of a straight cross between a plant homozygous for its presence with one homozygous for its absence. There are probably also some modifying genes which, in some stocks, allow only traces of HCN to be given off. The variant yielding HCN is morphologically indistinguishable from that which is free from the glucoside.

Rosaceae

RUBUS IDAEUS (raspberry). The raspberry occurs wild in woods and thickets in most of the British vice-counties, especially on lime-containing soils. It tends to multiply vegetatively by new canes which often arise from the roots so that aberrant variants may form extensive clones (populations produced by vegetative multiplication). An interesting sex variant with male flowers has been received from near Newbury and retained its characters unchanged in the experimental ground for a decade. Besides having flowers producing good stamens but no functional carpels, and therefore setting no fruits or seeds, it has rather small leaves with obtuse leaflets which are most often more or less united into a lobed but simple leaf. This combination of female sterility and peculiarities of leaf morphology has appeared also in some cultivated raspberries. The basic chromosome number for the genus is $n = 7$, and in the raspberries diploid, triploid, and tetraploid variants occur, but the sexual differences are not correlated with differences in the chromosomes. In addition to this male variant, and the usual hermaphrodite, female and neuter variants are also recorded. Genetical investigations show that two factors are concerned with sex : FM is hermaphrodite ; Fm is female ; fM is male ; and fm is neuter. The segregation with respect to the suppression of male organs is not so sharply discontinuous as that of the female organs. It is of interest to recall that sex behaves genetically in *Ranunculus acris* and *Rubus idaeus* in a very similar manner, and some botanists believe that the Ranunculaceae and Rosaceae may be closely related.

The colour of the spines is correlated with the colour of the fruits. Red- and tinged-spined plants have red fruits; green-spined plants have yellow fruits. PT gives red spines and fruits; T, tinged spines and red fruits; Ptt and ptt, green spines and yellow fruits. P intensifies the colour. The spine colour of P and pt is green, but it is probable that the colour of the fruits of Ptt is distinguishable from pptt, those of the former having a very faint tinge of red when fully matured, whilst those of the latter are a clear yellow. Hairiness is dominant to the sub-glabrous condition.

RUBUS FRUTICOSUS, in the broad sense (blackberry). The blackberries are notoriously very variable and exceedingly difficult to classify into species that are at all equivalent to the species of many other genera. Recent research has done much to explain these facts. The complexity of the group, in structure and behaviour, is correlated with hybridization, the occurrence of polyploidy, the formation of seeds by sexual and asexual processes, segregation even with asexual seed production, and vegetative propagation. Diploid species always behave sexually. Occasional unreduced germ-cells occur and take part in fertilization, giving rise to polyploid plants in which reproduction may be entirely sexual, entirely non-sexual, or partly sexual and partly non-sexual. Polyploid species vary in the degree to which asexual reproduction is developed, and a particular species may show a variation in reproductive behaviour depending on the species used as pollen-parent in cross-pollination. Segregation has been found to occur within non-sexual offspring. Development from a purely vegetative embryo-sac is the common method of non-sexual reproduction. Where an embryo-sac of structurally normal type is formed, segregation may occur as a result of crossing-over in a diploid egg or by the production of an embryo from a haploid egg or a haploid nucleus other than the egg. Reproduction by seeds (or, more generally, what are structurally organs of sexual reproduction) formed without fertilization is termed apomixis. Most often such a method of reproduction allows of no genetical variation and of no segregation or recombination of characters.

Thus it is, from the standpoint of evolution and natural selection, on a par with vegetative multiplication and the apomictic offspring are genetically no more than a clone. That segregation can sometimes occur with apomixis makes an important exception to the generalization that apomixis closes the road to further evolution through re-combination of genes. With such varied behaviour in reproduction and multiplication one would expect, as one finds in nature, great variation and the absence of clear-cut systematic classes based on a high correlation of many characters. There is considerable evidence that hybridization, some kinds of polyploidy, and apomixis are frequently associated.

GEUM URBANUM (common avens) and *G. rivale* (water avens) (Pl. 41, p. 206) are two widely spread species of very different appearance. *G. urbanum* is common in the south of Britain and in Ireland but becomes scarce towards the north, while almost the reverse is true of *G. rivale.* Typically, the habitats of the two are quite different. *G. urbanum* is found under hedges, on roadsides, in dry brushwood, and at the margin of dry woods, or in woodland clearings. *G. rivale* is a plant of marshy meadows, damp woods, and copses. Occasionally the two species meet and a hybrid swarm may then result. The F_1 hybrid between the two species is not exactly intermediate between them but shows on balance a greater resemblance to *G. rivale.* It is self-fertile and in the F_2 generation segregation is well-marked. No true-breeding types were extracted from the offspring of the straight cross. The F_1 has been called × *G. intermedium* (not a particularly happy name). Back-crosses with the parents gave different results according to the parent supplying the double dose. Thus, *intermedium* × *urbanum* gives offspring more like *G. urbanum,* but on selfing these there is very considerable segregation. On the other hand, *intermedium* × *rivale* gives offspring very like *G. rivale* and on selfing these very little segregation takes place ; in fact, unless very carefully examined, the plants might be mistaken for *G. rivale.* In one population studied in the wild in Berkshire, this *intermedium* × *rivale* back-cross appeared to have become stabilized and to be very largely replacing *G. rivale.*

POTENTILLA ARGENTEA. A brief mention may be made of this plant which occurs rather irregularly in Britain in gravelly pastures and on roadsides. It has been shown that there are many cytological variants within the species and these may be diploid ($2n = 14$), tetraploid

(2n = 28), pentaploid (2n = 35), hexaploid (2n = 42), or octoploid (2n = 56). There is further a considerable variation in sexual behaviour, from normal fertilization to complete apomixis. More or less sexual strains can be used to produce hybrids both between different variants and with other species. It is probable that apomixis in *Potentilla* is controlled by multiple factors. The balance between these factors may easily be broken by hybridization if one of the parents is more or less sexual.

ALCHEMILLA spp. (lady's mantle). It has long been known that plants of the aggregate species *Alchemilla vulgaris* reproduce apomictically. More recently it has been shown that the so-called microspecies are groups of apomicts which themselves show constant characters when grown under conditions as uniform as possible whether by cloning or from seed. Further, these genetically distinct types composing the taxonomic microspecies occupy different habitats and so enable a microspecies to have an extended range. In other words, the apomicts behave as ecotypes and the term agamotype has been proposed for them. It has also been found that the small annual *A. arvensis* is apomictic. As in the *A. vulgaris* apomicts, the embryo-sacs have diploid nuclei, owing to failure to complete a reduction in the number of chromosomes. Pollen develops normally and the pollen-grains can germinate on the stigma. The diploid chromosome number is 48. *A. arvensis* has an extensive geographical range, but shows very little genetical variation though it is highly plastic with varying environmental conditions. It would seem that its apomictic method of reproduction is of considerable age.

ROSA spp. (dog-roses) (Pl. 42, p. 207). Like the brambles and the hawkweeds, the roses are regarded as an exceedingly difficult group to classify. Whether recent researches by modern methods will enable more satisfactory systems to be made remains uncertain, but they are certainly explaining underlying causes of the strange mixtures of characters and peculiarities of behaviour in members of the genus. It is not easy briefly to summarize these researches, partly because authors are not always in agreement with one another, and sometimes not even with their own earlier conclusions, and partly because some of the most recent important research has been done on Scandinavian roses and it is necessary to be cautious in applying studies on foreign material to British populations.

Eric Hosking

MEADOW SAXIFRAGE, *Saxifraga granulata* (Saxifragaceae). Needham Market, Suffolk. April 1945

Brian Perkins

KNAPWEED, *Centaurea nigra eradiata* (Compositae)
Plant taken from near Leatherhead, Surrey. September 1946

The basic chromosome number in *Rosa* is $x=7$. There is no doubt that hybridization has been, and very probably still is, rife, at least between a large number of the microspecies. Some writers have accepted " the hybrid origin of practically every British rose " as a fair generalization. With this hybridization there is very considerable polyploidy, and diploid, tetraploid, pentaploid, and hexaploid British roses have been recorded. *Rosa arvensis*, one of the most easily recognized and clear-cut species (if the term be applicable) of British roses, is diploid. In all the diploids examined, reduction of chromosome number was normal and, presumably, from the evidence available, there is normal sexual reproduction. All other British roses, so far as they have been examined, have abnormal partial reductions, involving 14 to 28 chromosomes and resulting in different chromosome numbers in the pollen-grains and egg nuclei respectively. Many British roses are pentaploid ($2n=35$, i.e. with 5 sets of the basic $x=7$), so we may take one of these as an example. Irregularities occur, especially in the formation of pollen-grains, but, in general, such irregularities result in collapse and non-functioning of a proportion of the pollen-grains. Examination of a sample of dog-rose pollen under the compound microscope nearly always shows a number of " bad " grains. Since these pollen-grains with unusual chromosome contents do not generally function we can concentrate on what is the usual behaviour—usual, that is, in the pentaploid roses, but unusual for plants as a whole. There are 35 somatic chromosomes. In the reduction division leading to the formation of male gametes, pairing occurs to produce 7 bivalents, while 21 chromosomes remain unpaired, though all are orientated on the equator of the spindle. The bivalents separate and move to the poles so that 7 chromosomes enter each pollen nucleus. The 21 univalents, though they first divide, get " left behind " in the final reduction phase, so that the pollen-grains, even in the pentaploids, are normal for the genus in their chromosome number. In the developing ovule there is a different behaviour. 14 chromosomes again pair to produce 7 bivalents which alone orientate themselves on the equator of the spindle and behave normally, in that 7 of the chromatids enter the constitution of the functioning embryo-sac nucleus (and eventually the egg nucleus). The 21 unpaired (univalent) chromosomes all remain in the chalazal part of the cell, where they are joined by 7 of the chromatids from the bivalents. Thus the functioning embryo-sac nucleus (and eventually the egg nucleus) comes to contain 7+21 = 28 chromosomes. We thus have this peculiar scheme :

	male side		*female side*
Somatic chromosomes	35		35
Pairing	(7+7)+21		(7+7)+21
Reduction	7 (lost 21)		7+21
Gametes	7 ↘		↙ 28
Fertilization & zygote formation		35	
Somatic chromosomes	35 ↙		↘ 35

It should be stated clearly that it is not yet proved that this scheme holds for all British pentaploid (and with mere differences in the numbers of univalents for other British polyploid) roses. The evidence, however, seems to be accumulating that this is the explanation of the diversity mixed with frequent true-breeding of British roses of the polyploid series, i.e. of a very large proportion of our British rose flora.

There remains one still controversial question on which there has been a considerable diversity of opinion. Earlier cytologists studying roses were impressed by the appearance, in a predominant or complete manner, of characters of the mother plant and concluded that roses were largely apomictic in their setting of seed. There is a tendency now to regard apomixis as exceptional, in many of the roses at least, even if it occurs at all. The matter cannot be considered as settled, especially for the British roses. An attempt has been made here to summarize researches which are being continued, and tentative conclusions may have to be modified or rejected in the future. The production of chromosomally different male and female gametes with regular fertilization does seem to fit the very wide range of facts as at present known, and if further research establishes the scheme as true and of wide application we have an extremely interesting example of the evolution of a special reproductive behaviour. Its advantages are obvious for it allows reasonable constancy with ample variation, the advantages of polyploidy with fertility, and reasonable chances of hybridization and recombination of genes and characters.

One other matter may be mentioned here. Rose-hips are a well-known source of vitamin C (ascorbic acid). There are considerable differences in the vitamin C content in different roses and attempts have been made to correlate these differences with differences in cytology, distribution, and degree of fertility. There is some evidence that, at least within the same group for Europe as a whole, higher polyploidy and more northern range are correlated with higher vitamin C content. Another correlation, which it has been claimed can be established for a number of Scandinavian roses, is between degree of fertility and vitamin C content. Rose-hips, that is the fleshy outer part of the fruit, will develop fully if only one or two or even none of the nutlets inside produce mature embryos. It is stated that in some roses the vitamin C content of the hip is lower the greater the number of nutlets that develop per hip. No marked negative correlation has, however, been found for this in British dog-roses. For the genus as a whole, species ripening in August, when grown in the London area, are richer in vitamin C than those ripening either later or earlier. There is also a close correlation between the systematic position and vitamin C. content. For the British roses it is probable that the production of vitamin C in the hips is mainly under genetic control.

SAXIFRAGACEAE

SAXIFRAGA spp. (saxifrages). Breeding work has been done on British material of *Saxifraga granulata* (Pl. 43, p. 210) and a number of other species. The results, which are rather complicated, have not all been published. In *S. granulata* certain variations in petal shape have been shown to have a genetical basis. The occurrence is recorded of female plants, in which, as in so many other species showing gynodioecism (i.e. having female and hermaphrodite variants), the petals are reduced in size. Most interesting results were obtained by crossing *S. granulata* with other species. Thus, when it was crossed with *S. rosacea* from Western Ireland, a uniform F_1 family was obtained in which the habit was very similar to that of *S. granulata*. Sepal and petal shapes were intermediate, while fruit shape and seed size were more like these characters in *S. rosacea*. In F_2 families there were differences. In five families there was marked segregation into grandparental and intermediate characters, i.e. there was expected, though complicated, Mendelian behaviour. In twenty-nine families (with a total of 4,068 plants of F_2, F_3, and back-cross origins) there was no segregation. It had to be accepted that a new true-breeding kind had been created and the name *Saxifraga potternensis* was suggested for it.

The chromosomes in *Saxifraga* are not easy to study, as they are numerous and small, and it has proved somewhat difficult to obtain satisfactory preparations of them. The latest investigations gave the following results: *S. granulata* $2n=48–49$; *S. rosacea* $2n=64$; F_1 hybrid $2n=56$; F_2 (*S. potternensis*) $2n=$about 80. From these findings it is concluded that, relative to the basic number, $x=8$, *S. granulata* is hexaploid, *S. rosacea* octoploid, the F_1 hybrid heptaploid, and *S. potternensis* approximately decaploid.

S. potternensis showed a very close resemblance to *S. granulata* in vegetative characters and behaviour. It was only when it was back-crossed to *S. rosacea*, i.e. when apparently two "doses" of *S. rosacea* were added to one "dose" of *S. granulata*, that *S. rosacea* characters of habit and vegetative behaviour were evident. This does not, however, prove by itself that one set of chromosomes of *S. granulata* is, for expression of specific vegetative characters, the equivalent of two sets of *S. rosacea*, since both *S. granulata* and *S. rosacea* are polyploid and the still higher polyploidy of *S. potternensis* may be due to a relatively greater increase in sets of *S. granulata* than *S. rosacea* chromosomes.

In addition to the above successful crossings, hybrids have been obtained between *Saxifraga tridactylites* and *S. granulata*, *S. granulata* and *S. hypnoides*, and *S. tridactylites* and *S. hypnoides*. The three species involved in these crosses belong to three different taxonomic sections of the genus. *S. tridactylites* is an ephemeral annual, the other two are perennials. Using *S. tridactylites* as the ovule and *S. granulata* as the pollen parent, a small F_1 family of four plants was obtained. Selfing one of these a family of 200 plants was raised. Both F_1 and F_2 families were indistinguishable from *S. granulata*, a result still awaiting a cytogenetical explanation. Another selfing and a crossing between two F_1 plants gave the same result. The cross *S. tridactylites* (ovule parent) × *S. hypnoides* (pollen parent) gave a family of 10 plants somewhat intermediate in some characters but in the perennial habit and most other characters nearer to *S. hypnoides* (the pollen parent) than to *S. tridactylites*. These plants were completely sterile, with their own pollen and in back-crosses with either parent. This hybrid has been recorded wild on the western face of Ingleborough and named *S. farreri*. When *S. granulata* (ovule parent) was crossed with pollen from *S. hypnoides* a family of 40 plants was raised all very similar in resembling the *S. granulata* parent more closely than *S. hypnoides*. One of these plants selfed gave a family of 96 plants and from these further families were obtained by selfings and crossing two together. Throughout these families there was no segregation towards the specific characters of *S. hypnoides*, except for the occurrence of simple leaves and variation in petal shape. Again the overwhelming influence of *S. granulata* was shown.

Lythraceae

LYTHRUM SALICARIA (purple loosestrife). This species has been investigated in some detail, but mainly with the hope of elucidating one phenomenon, namely that of tristyly. A certain number of species, belonging to different families, have, in different individual plants, flowers with different lengths of the styles and, associated with these differences, different lengths or positions of the stamens. This phenomenon is known as heterostyly and is considered to be a structural device which encourages cross-pollination. The facts recorded here for the purple loosestrife may be compared with the simpler phenomena found in the primrose (see p. 224). In the purple loosestrife there are three kinds of flowers, every plant having flowers of one kind only. These kinds are conveniently named, according to the length of the style: long, mid, and short, but, as the accompanying diagram shows, different

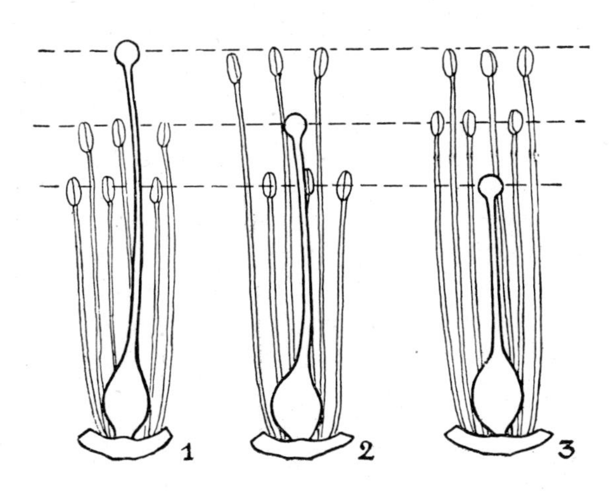

filament lengths are associated with different style lengths. Pollination, and resultant fertilization, between anthers and stigma at the same level is termed " legitimate " mating. Breeding results have shown definitely that tristyly in purple loosestrife has a genetic basis. " Long " is a full recessive, since long-styled plants on selfing yield only similar offspring. " Mid " is dominant over " long," and " short " over " mid " and " long." There are, however, some complications. Thus, some mids, though behaving almost as if homozygous, give a small percentage of longs when crossed with a long. Several theories have been proposed to explain these results and it now seems most probable that they are due to a gene being represented a number of times, owing to polyploidy, in as many similar but separate chromosomes. This type of inheritance is known as polysomic and quite obviously can result in ratios which will often, but not always, be different from those obtained when only one pair of chromosomes is involved.

Generally, all three kinds of plants, those with long, mid, and short flowers respectively, occur in any fair-sized populations. Actual proportions differ greatly in the counts recorded. Sometimes mids are below the 33·3 per cent that would be expected in exactly balanced populations. Mids are said for some stocks to yield most seeds per capsule but are less efficient as pollen parents than the longs and shorts, though in one population in culture mid plants bore some 16·1 per cent fewer viable seeds than longs and shorts. In other counts, shorts were below the 33·3 per cent of a balanced population. It may be that there are definite differences between different stocks in this respect, or that habitat differences react differently on longs, mids, and shorts. Besides the differences in lengths of styles and stamens, the three kinds of flowers have each two kinds of pollen. The grains differ in size according to the stamen length : long pollen is largest and mid is larger than short. The greatest divergence is between the long-stamen pollen of the short-style plants and the short-stamen pollen of the long-styled plants, the ratio being 100 : 56·6 (as based on diameters). There are also differences in pollen colour. Generally, mid- and short-stamen pollen is pure yellow, while long-stamen pollen is of a varying intensity of green. Exceptions and variations have, however, been recorded.

In the British Isles, it has generally been stated that there is no difficulty in classifying plants of purple loosestrife into one of the three groups : long, mid, or short. In a cultivated population in the New York Botanical Garden, however, a new type appeared with the set of longer stamens almost equal in length to the style, and this length about midway between the two lengths characteristic of the mid and long lengths as seen in long-styled and in mid-styled plants. The plant set abundant seed after self-pollination and 122 plants were raised from this seed. Of these, 60 were of new-type like the parent, 9 were long-styled, 52 intermediate between the new-type and long-styled, and 1 was mid-styled. Such aberrant types should be investigated in detail if found in Britain.

ONAGRACEAE

EPILOBIUM spp. (willow-herbs). Some species of *Epilobium* hybridize very readily and wild hybrids are frequently recorded in Britain. Nevertheless, the species keep distinct as a whole. Extensive experiments have been made under control and these throw some light on the problem. In crosses between different species, and sometimes in crosses between different stocks of the same species (as in *E. parviflorum* and *E. hirsutum*), a proportion or all of the offspring are dwarfed, contorted, or variously malformed. Many are strikingly different in appearance, which might be described as pathological, from the parents, and die before flowering. Some are lethal at very early stages. These dwarfed, contorted individuals stand out strikingly in cultivated families and seem to have factors which have checked growth or normal development. The particularly interesting point has emerged that reciprocal crosses often give quite different results and that factors responsible for development are present in the cytoplasm, i.e. outside the nucleus, and are therefore not situated in the chromosomes. The most recent research indicates that there are also nuclear genes which play a part in the development of habit and that there is a complex interaction between nuclear genes, cytoplasmic factors, and environmental conditions, which together are responsible for the normal or abnormal development.

CUCURBITACEAE

BRYONIA DIOICA (the white bryony) is our only native British plant of the marrow family (Cucurbitaceae). It is normally dioecious but a hermaphrodite variant has been grown at Kew from seed collected from a garden near Southport. This comes true from seed and its cytogenetics is being studied. Some breeding research has been done with dioecious individuals, varying in other than sex characters. The absence of a waxy bloom from the ripe berries behaves as a simple dominant to the presence of waxy bloom. Co-operation of two factors accounts for the results obtained by crossing two varieties with two-carpellary and three-carpellary flowers respectively, as shown by the number of stigma-lobes and placentae in the ovary. Capacity to increase the number of vascular bundles in the stem beyond ten is a simple dominant to absence of such capacity. Differences in habit and foliage have a genetic basis but are evidently complex in origin, though two factors may account for differences in leaf-shape between two of the varieties that were studied.

Valerianaceae

VALERIANA OFFICINALIS (common valerian). Some very interesting results have recently been obtained from researches on this species. Two chromosome numbers occur in British representatives: the one is tetraploid (2n = 28) and the other octoploid (2n = 56). There are no sharply marked constant differences in external characters between the two polyploids; each includes a number of more or less distinct variants, differing, probably genetically, in shape of leaflets, toothing of leaflet margins, pubescence of stem, presence of anthocyanin, and other characters. These variants, with environmentally determined plasticity, cause intergrading in gross morphological features between the two polyploid types. Field studies and the examination of herbarium specimens show that the ranges, in Great Britain, of the tetraploid and octoploid are different: the former, in general terms, is southern and the latter northern and, with some exceptions, a line from the Wash to the head of the Severn estuary marks the northern limit of the tetraploids. In some southern and midland counties, for example in Gloucestershire, the ranges overlap—but there are then different ecological preferences controlling local distribution. The tetraploids occur especially on Jurassic (oolitic) limestones and on the Chalk, and sometimes, in the south-west, on Carboniferous Limestone, and are xerophytic, growing in a relatively dry soil, in hilly regions, on slopes and in dry valleys. The octoploids occurring within the area of the tetraploids grow at lower altitudes in moist soil, along river-sides and

PLATE 45

F. Ballard

KNAPWEED, *Centaurea nigra radiata* (Compositae). Paignton, Devonshire. August 1946

Brian Perkins

GREATER KNAPWEED, *Centaurea scabiosa* (Compositae), Banstead Downs, Surrey, August 1946

on ditch banks. Farther north, however, the octoploids are represented by variants growing in both dry and wet places. It would appear that the octoploid, presumably a younger type, has a greater toleration range and has thus been able to increase its range more than has the tetraploid.

Compositae

SENECIO VULGARIS (common groundsel). This is one of our most ubiquitous weeds of arable land and waste places and can be found in flower, in the south of England, every month of the year. It is highly plastic and a number of genetically determined variants have also been studied. It is probable that the number of nodes has a genetic basis with four pairs of characters involved. Other factors which lead to segregation of characters in F_2 and, except for some linkage, give expected results on the basis of simple Mendelian behaviour, can be conveniently tabulated thus:

Indumentum factors

Main factor for hairiness:	H = hairy, h = glabrous.
Factor modifying hairiness:	Y = depression of hairiness, y = allows full development of hairiness (if H be present).
Factor affecting stimulation of H by R (see below):	Z = reduction of hairiness in R plants, z = normal hairiness in R plants.

Stem colour factor

Factor for red and green colour:	G = red, g = green.

Leaf colour factor

Factor for intensity of green:	L = dark green, l = yellow green.

Floret factors

Factor for ray florets:	R = radiate, r = non-radiate.
Factor for floret colour:	C = yellow, c = cream.
Factor influencing cream colour:	Xc = yellow, xc = cream.
Factor influencing ray character:	F = normal rays, f = fimbriate rays.

In families from certain crosses between variants (in which both parents were normal green) albino seedlings appeared. The albinism was interpreted as a double recessive—one parent was AAbb and the other aaBB. It is probable that, in addition, there are other genetical types of chlorophyll abnormalities within the species.

CENTAUREA spp. (knapweeds). These plants are common in grass verges, in some old permanent pastures, and on some grassy heaths and downlands on a variety of soils. The species in Britain may be divided into two groups (apart from casuals and obvious introductions): the *C. scabiosa* and the *C. nigra-jacea* groups.

CENTAUREA SCABIOSA (Pl. 46, p. 129) shows considerable variation in leaf-shape and in floret colour and shape but it has not been successfully crossed with any member of the *C. nigra-jacea* group, though it has yielded hybrids with the yellow-flowered European *C. collina*. A series of genes influencing flower colour within the one species *C. scabiosa* has been symbolized as follows:

A or B gives marguerite yellow.
AB give dull magenta purple.
ABC ,, tourmaline pink.
ABD ,, mallow pink.
ABDE ,, dull dark purple.

Within the *C. nigra-jacea* group three kinds of variation have been recognized.

1. *Habitat fluctuation.* This is not of great importance. Thus on the different soils of the Transplant Experiments of the British Ecological Society at Potterne the knapweed used showed very little variation in morphological characters or observed behaviour compared with other species, such as *Plantago major*, on the same soils. Dwarfed forms occur in exposed situations, such as some cliffs and chalk downs, and grazed pastures. These have not been widely tested but some are only habitat fluctuations while others may be ecotypes with different gene constitutions.

2. *Intra-specific variation* with a genetic basis is represented by such characters as radiate (Pl. 45, p. 218), semi-radiate, or eradiate (Pl. 44, p. 211) capitula; long, medium, or short florets; sex characters (in part correlated with floret size); floret-colour variations; leaf-shape variations (in basal and cauline leaves); habit (to a certain extent); and precocity or lateness of flowering. These characters segregate from crosses in various combinations.

3. *Variations due to inter-specific hybridization.* These are very numerous, affect most parts of the plant, and are best detected as such by a preliminary examination of the phyllaries (the bracts around the heads

of florets). In particular *C. jacea* crossed with microspecies of *C. nigra* gives rise to mixed hybrid-swarms due to segregations and back-crossings. Whether any of the knapweeds are strictly native to Britain is an open question. Whether naturally or through the agency of man, *C. nigra* was probably the first to invade these islands and has spread widely, especially along road- and path-verges. Unrayed variants of this are the usual knapweeds in the north, east, and west of Britain. In the south and south midlands there is also *C. nemoralis* which is very often rayed. Locally, again mainly in the south and south midlands, *C. jacea* has appeared, almost certainly as an introduction with crops, in chicken food, etc. Nearly all knapweeds have a high degree of self-sterility and are favourite plants for the visits of bees. Hence cross-pollination and, within the *nigra-jacea* group, cross-fertilization readily occurs. Wherever *C. jacea* contacts a population of either *C. nigra* or *C. nemoralis*, crossing occurs again and again. Since *C. jacea* is usually present at its introduction as few individuals while the other knapweeds may be represented by hundreds or thousands within pollinating distance, the tendency is for *C. jacea* to be "swamped" and after some years its previous presence can only be detected by character combinations and individual variations. It was not till a long series of breeding experiments had been completed that a correct analysis of naturally occurring populations became possible. Such hybrid swarms have now been studied in Berkshire, Oxfordshire, and Surrey.

Not only do breeding experiments and field studies agree with the above outline of the history and behaviour of knapweeds in Britain but the cytological findings afford a further explanation. *Centaurea scabiosa* has $2n=20$; *C. jacea*, *C. nemoralis*, *C. nigra*, and their variants have $2n=44$. On the basis of chromosome numbers one would not expect *C. scabiosa* to cross with any member of the *C. nigra-jacea* group, while there is no obvious reason, from the standpoint of chromosome numbers, why these latter should not cross with one another and yield fertile offspring.

TARAXACUM spp. (dandelions) (Pl. 47, p. 222). "The dandelion" is a misnomer. Actually in Britain there are a very large number of biotypes which come true from seed. These differ one from another very considerably in characters of habit, foliage, inflorescences, flowers, and fruits. They can only be adequately studied if the scheme of investigation includes cultural experiments and especially growing rows of biotypes under conditions kept as uniform as possible. The reason for this statement is that a *Taraxacum* plant is extremely plastic in habit and leaf form, these characters being influenced especially by the supply of food, light, and moisture, and by the age of the individual. A succession of leaf-forms marks growth from the seedling stage to the adult condition and there is also a seasonal sequence of leaf-forms. The series of leaf-shapes is not the same for all biotypes but generally there is fair stabilization from the second year till the plant gets old and tends to split up vegetatively at about ground level, often about the fifth year. These leaf-shape differences mainly concern degree and kinds of lobes and teeth. At any age of the plant, moisture is an important factor in controlling leaf-shape : kept in dry in the shade the leaves become shorter and more dissected ; in the shade in a saturated atmosphere only entire leaves develop. In spite of such, and much more, plasticity, there is no doubt that many differences are due to plants belonging to different genotypes. Dandelions grown in an experimental ground where conditions are kept as nearly to one standard as is possible in out-of-doors experiments, show, just before flowering, remarkable agreement in vegetative characters between plants grown from seeds from one head, while rows from seeds of different parentage often contrast markedly. In scores of experiments with British dandelions, no segregation has been detected up to the present in " one head, one row " cultures.

Can we explain this diversity of biotypes combined with constancy in reproduction from seed ? To some extent we can and the facts now known enable us to state the unsolved problems rather precisely.

Firstly, all the British dandelions so far tested have been found to be apomictic : viable seed is set without fertilization. Secondly, all the British dandelions so far examined cytologically are triploid ($2n = 24$), the basic number for the genus being $x = 8$. Thirdly, there is evidence of a fairly high correlation between habitat preferences and some, but not all, sections and biotypes. Populations in natural or semi-natural habitats are much more uniform (in biotype composition) than in disturbed, waste, or ruderal areas. A sand-dune area, a chalk grassland, or a limestone ridge has a much less diversified dandelion population than many a much smaller area of broken field, waste ground, or newly made-up verge or embankment.

Some of these facts may be discussed in a little more detail. Dande-

PLATE 47

John Markham

DANDELIONS, *Taraxacum officinale*, apomicts (Compositae). From roadside, West Pennard, near Wells, Somerset. April 1946

John Markham

Rice Grass, *Spartina townsendii* (Gramineae). Chichester Sound, Sussex. November, 1946

lions set viable seeds if heads, with stalks, are gathered before the fruits are ripe and kept in water indoors. This is a great convenience in experimental research because as soon as the pappus unfolds a fruit will blow away in even a light wind. It is a simple matter, at least after a little experience, to slice off, with a safety-razor blade, the top parts of the florets to the level of the top of the ovaries in an inflorescence bud before this opens. By this operation anthers and stigmata are removed and, if plants so treated are kept moist by being covered with a bell-jar and shaded, though fertilization is impossible, viable seeds are still set. Such seeds will germinate immediately but lose their vitality generally within 2 to 3 years. In the genus as a whole, there are other polyploids besides triploids, thus tetraploids and pentaploids are recorded and quite probably some British dandelions have $2n=32$ or $2n=40$. Moreover, dandelions reproducing by normal sexual means, with pollination and fertilization, occur, though they have not yet been recorded for the British flora. Such plants producing viable seed following fertilization have been grown at Kew from seed received from Japan and from Russia. The much-discussed *Taraxacum kok-saghyz*, a native of South Russia, is diploid ($2n=16$) and reproduces sexually. It is this plant which is grown for its latex from which rubber is produced. A sexual diploid has also been recorded from Scandinavia. It is of interest to note that some at least of these diploid dandelions can readily be crossed with one another and produce fertile offspring. In spite of the fact that seed is set without fertilization in the triploid dandelions, they retain their elaborate mechanisms for pollination and the only sign of reduction is the often poor development of pollen.

The available evidence suggests that normal sexuality and resultant seed-setting were formerly of general occurrence in the genus, that hybridization combined with the production of polyploids gave rise to a number of phenotypes which were largely apomictic, and that subsequent mutation has been, and perhaps still is, the origin of many of the numerous biotypes now known especially from N.W. Europe, including the British Isles. It is highly probable that somatic (including zygotic) mutations occur now and again and give rise to new biotypes, which, retaining apomixis, are already stable in the sense of coming true from seed. Thus a " quilled " dandelion was found in a field near Manchester. In this the normally strap-shaped corollas are fused into narrow, closed tubes in such a way that only a pale under-surface of the petals was exposed to view. This variant was reproduced by seeds, and by decapitating was shown to be apomictic. Parallel variants have been recorded on the European continent. The chances of mutations occurring are high because of the large numbers of seeds produced and the high percentage of germination. Over 23,000 seeds were produced by one plant in one season, though, it is acknowledged, this plant was growing under almost the best possible conditions.

The distribution of the biotypes (many, at least, of them apomicts) in the British Isles awaits detailed study. Dandelion populations in the north of Scotland are, in the examples examined in the field and in the experimental ground, different from those in the south of England and there are differences between English populations. Moreover, there is evidence of natural selection of biotypes. It has proved, as with some other common species, difficult to grow Icelandic dandelions successfully at Kew with conditions under which British stocks thrive and even become a local nuisance! The "major species" are fairly well characterized morphologically and have distinct habitat preferences. Thus the common dandelion (*T. officinale*) flourishes best in open communities, the red-fruited dandelion (*T. laevigatum*) in short grassland, and the marsh dandelion (*T. palustre*) in wet habitats.

PRIMULACEAE

PRIMULA VULGARIS (primrose) (Pl. XXIV, p. 251). This species shows very typical heterostyly with flowers of two kinds, but on any one plant of one kind only. In the "pin-eyed" flower the style is long so that the stigma is at or very slightly above the mouth of the corolla tube and the five anthers are near the middle of the tube. In the "thrum-eyed" flower the style is short so that the stigma is in position at about the middle of the corolla tube while the anthers are near the top of the tube, practically at its mouth. The long-styled plant has a more globular and rougher stigma, smaller oblong pollen-grains, the upper half of the corolla tube more expanded, and larger ovules, and produces fewer seeds per fruit. The short-styled plant has a smooth depressed stigma, larger spherical pollen-grains, the corolla tube with uniform diameter except close to the upper end, somewhat smaller ovules, and more seeds per fruit. Cross-pollination between the two kinds is usually necessary to ensure anything but a very small quantity of seed-setting.

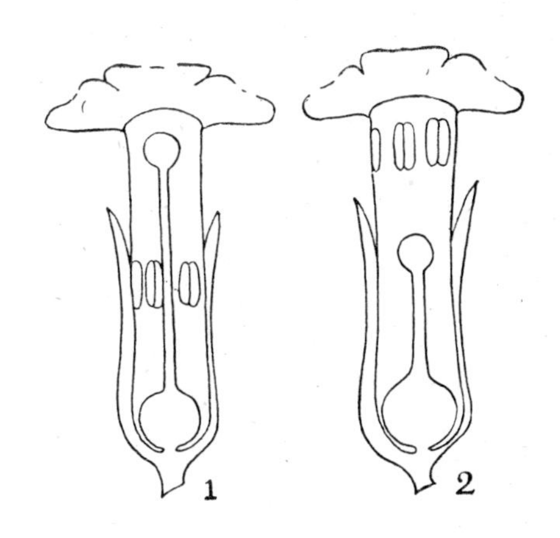

In wild populations in Dorset and Surrey counts of over 1,000 plants gave almost exactly half of each kind as composing the random samples. Some populations from Somerset gave very different results : (1) 102 pin : 11 thrum : 210 long homostyle ; (2) 145 pin : 15 thrum : 468 long homostyle ; (3) 152 pin : 103 thrum : 177 long homostyle ; (4) 46 pin : 75 thrum : 7 long homostyle ; (5) 41 pin : 40 thrum : 2 long homostyle. The long-homostyle flowers had the stamens in the thrum position at the top of the corolla tube but the style was long and the stigma in the pin position. Experiments showed that long homostyle was quite self-fertile but no account of its genetics appears to have been published. In ordinary heterostyled stocks long-style is recessive to short-style. Rather aberrant F_2 ratios are sometimes obtained and it remains uncertain whether a single pair of genes is involved or whether several determine heterostyly, or whether there is some differential viability—plants of certain composition sometimes failing to establish themselves so readily as those with other gene combinations.

A considerable number of " abnormalities " have been recorded in the primrose. In one there were enations from the corolla lobes. These took the form of a pair of petaloid outgrowths or flaps on the upper side of each lobe of the corolla. They showed inverted orientation relative to the corolla lobes, that is the lower epidermis (outermost layer of cells) of the enation was continuous with the upper epidermis of the lobe, and the vascular bundles had their parts reversed. The enated condition was recessive to the normal in the F_1 families resulting from a cross, and segregation occurred in F_2 families with what probably represented a 3 normal : 1 enated ratio. Another very striking abnormality involved the stamens and gynoecium, various strange-looking metamorphosed structures, showing combinations of imperfect petal, stamen, and carpel characters occurring inside the flower and resulting in distortion of calyx and corolla. This condition also was, in F_1, recessive to the normal on crossing, and segregation occurred in F_2 with approximation to a 3 normal : 1 metamorphosed ratio.

Coloured primroses are often grown in gardens. In some countries, as in eastern Bulgaria, a red primrose occurs to the exclusion of the yellow. Examples of coloured primroses from a population growing near Tenby, Pembrokeshire, have been studied genetically. A considerable range of colours was obtained. The yellow colour of the ordinary primrose is a flavone colour associated with plastids, while reds and blues are colours from anthocyanins in solution in the cell sap. A factor Y gives yellow flavone colour and, in the Pembrokeshire plants and their offspring, this was always present, giving a yellow background against which the anthocyanin colours were superimposed. Factors modifying the action of Y occur in some primroses. A factor B results in the reproduction of anthocyanin. When B is present another factor, S, causes reddening, in single or double dose. It and its allelomorph possibly act through selective permeability: in the reds (with S) strongly basic ions being excluded from the cells, and in blue primroses (not present in the Pembrokeshire plants studied, but known in gardens) passing into them and modifying the red colour to blue. Another factor, N, acts as an intensifier, giving deeper and, on the whole, duller colours.

Scrophulariaceae

DIGITALIS PURPUREA (foxglove). A number of variants, some striking, are known to occur in this familiar species. Variations in flower colour are not infrequent, even in the wild, and some at least have a genetic basis. Thus, the following factors have been accepted:

- M giving magenta colour, and dominant to m, which gives a recessive white.
- D is a darkening factor, converting magenta to purple, but acting only when M is present.
- W is a dominant factor for white; in its presence the expression of colour due to M is inhibited so that the flowers are white. Thus there are both dominant and recessive whites, of different genetical constitution but phenotypically similar in general colour.

All flowers appear to be spotted. In the presence of the colour factor, M, spots are red: in the absence of M they are yellow. The

John Markham

LORDS AND LADIES, *Arum maculatum* (Araceae). Northaw, Hertfordshire. April 1945

PLATE XVIII

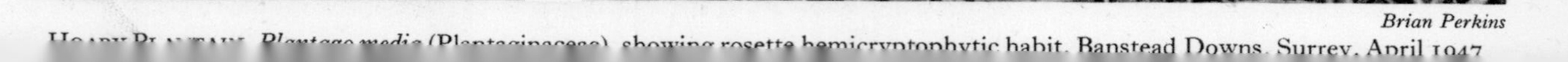

Brian Perkins

HOARY PLANTAIN, *Plantago media* (Plantaginaceae), showing rosette hemicryptophytic habit, Banstead Downs, Surrey, April 1947

presence of the dominant white factor, W, does not inhibit the expression of the colour factor M in regard to the spots (at least when W is present in single dose, Ww).

There are several floral abnormalities which crop up from time to time in foxglove populations. One of the best known is the formation of a large bell-shaped flower at the end of the raceme. This flower does not show the typical zygomorphy of the normal flowers and stands up more or less vertically. Its corolla may be 9-lobed, the anthers are up to 8, and the stigma has as many as 5 stigmatic surfaces. This monstrous condition is recessive to the normal. It is interesting to note that the lower normal flowers of a raceme ending in a monstrous (probably structurally composite) flower also carry the recessive allelomorph (alternative gene) for producing monstrous terminal flowers, since seeds from them, by self-pollination, produce plants having this structure.

In another abnormality, the corolla is split into separate members, some or all of which are converted into stamens. The result is a flower which is hardly recognizable as a foxglove at all. This variant also is recessive to the normal.

Two variants are known which differ in stem indumentum. In one the stem is grey and densely pubescent throughout and the leaves are very hairy (var. *pubescens*). In the other the stem is green, polished, and smooth from the base to the flowering region where it becomes pubescent, and the leaves are less hairy (var. *nudicaulis*). Each sets seed abundantly and breeds true when kept pure. The var. *nudicaulis* behaves as a Mendelian dominant when crossed with var. *pubescens*, and the hybrids segregate in a ratio of 3 *nudicaulis* : 1 *pubescens* when self-fertilized and 1 *nudicaulis* : 1 *pubescens* when back-crossed with the recessive var. *pubescens*.

Labiatae

GALEOPSIS TETRAHIT (common hemp-nettle). This annual occurs as a weed in cultivated and waste places and sometimes in half-shade in woods and hedges. A great deal of cytogenetical research has been carried out on this and other species of the genus in Sweden. Some of the results are of great evolutionary interest and are therefore mentioned here.

Crosses between *Galeopsis pubescens* and *G. speciosa*, both having the haploid chromosome number 8, gave highly sterile F_1 hybrids.

However, an F_2 generation of about 200 individuals was raised. One F_2 plant was found to be triploid and almost completely sterile. After pollination with pollen from *G. pubescens*, one of the grandparents, the triploid plant produced one viable seed. This seed germinated to produce a tetraploid plant (2n=32) with good fertility. Both the triploid F_2 and its tetraploid offspring were structurally indistinguishable from *G. tetrahit* and the tetraploid behaved like wild *G. tetrahit* genetically, which Linnean species also has 2n=32. Later investigations tend to confirm that *G. tetrahit* arose from a combination of *G. pubescens* and *G. speciosa* chromosome sets, and agree with the conclusion that the artificial *G. tetrahit* is the first instance of the synthesis of a wild previously recognized species by controlled hybridization. Artificial *G. tetrahit* is fertile with wild plants of *G. tetrahit* and differs essentially from this only in not being quite so constant. In all probability constancy would be obtained in time by selection.

Experiments have shown that there is intraspecific sterility between different stocks within natural *G. tetrahit.* Correlation was found between flower colour and speed of vegetative development and between flower colour and partial sterility. The sterility is caused by lethal factors which are linked to the flower colour factors. Hairiness is determined by two dominant polymeric factors, the glabrous condition being a double recessive. Flower colour is mainly conditioned by one factor-pair and three systems of multiple allelomorphs. Crosses in the self-fertile *G. tetrahit*, between different stocks which had been so in-bred as to be considered pure lines, resulted in hybrid vigour in the offspring. By colchicine treatment autopolyploids have been obtained in *G. tetrahit* with 2n= about 64.

LAMIUM spp. (deadnettles). Two series of chromosome numbers have been found in species of *Lamium.* In *L. maculatum*, *L. album*, *L. purpureum*, and *L. amplexicaule* the haploid number is n=9; in *L. hybridum* and *L. moluccellifolium* it is n=18, as it is also in *Galeobdolon luteum.*

The red-flowered annual deadnettles are of particular interest. *L. purpureum* and *L. amplexicaule* are the most distinct in morphological characters, while *L. hybridum* and *L. moluccellifolium* are in many respects more or less intermediate, though distinct from one another. Some botanists have suggested that they might be of hybrid origin, *L. purpureum* and *L. amplexicaule* being the putative parents. Genetical experiments furnish no proof that this hypothesis is true. A large number of reciprocal crosses between all four species entirely failed, with only one exception. *L. amplexicaule* × *L. hybridum*, and the reciprocal, yielded offspring, which, however, were entirely sterile. These hybrids were in some characters intermediate between those of the parents, but in other characters, and particularly those regarded as specifically separating *L. moluccellifolium* from *L. amplexicaule* and *L. hybridum*, they agreed with *L. moluccellifolium*. The experiments have been carried out twice by different investigators independently and using distinct stocks. It seems most likely that the two intermediate species have existed independently as long as have *L. purpureum* and *L. amplexicaule*, all four having been derived from a common ancestor in which a spontaneous tetraploidy started the development of *L. hybridum* and *L. moluccellifolium*, while *L. purpureum* and *L. amplexicaule* kept the original chromosome number during their differentiation.

An example of genetic inheritance of pollen colour is furnished by *L. hybridum*. The pollen is generally scarlet-red, as it is in *L. purpureum*, *L. moluccellifolium*, and *L. amplexicaule*. In *L. hybridum*, however, variants with citron-yellow-coloured pollen occur. These breed true for this character and give offspring with only yellow pollen when crossed together. The cross, yellow × red, showed in F_1 almost complete dominance of red, with all pollen-grains equally coloured. In F_2 there was monohybrid segregation, a summated ratio of 303 red : 118 yellow being obtained. In F_3 some red plants bred true and others segregated in approximately 3 : 1 ratios. The difference concerns only pollen colour, the flower colour being the same in all the stocks concerned and their offspring. There are differences in in-bred lines in leaf shape and other quantitative characters, with apparently complex segregations. A variant with corollas split down the front has been recorded.

L. purpureum (the common purple-flowered deadnettle) has variants differing in leaf shape and flower colour. A white-flowered variant was recessive to the red-flowered and in F_2 segregation occurred with 79 red : 28 white.

L. amplexicaule is also variable. The flowers are generally spotted inside but some plants have wholly red flowers. Spotted crossed with red gave spotted in F_1 and in F_2 a ratio of 64 spotted : 22 red. This species often has cleistogamic flowers, i.e. small flowers which do not open and in which pollination takes place in the closed buds. These closed (cleistogamic) flowers occur on plants which may also have open (chasmogamic) flowers, though some plants may produce only cleistogamic flowers. Certain strains can be distinguished with different proportions of cleistogamic flowers. Two such had averages of $42 \cdot 75 \pm 0 \cdot 69$ per cent and $16 \cdot 01 \pm 0 \cdot 60$ per cent cleistogamic flowers respectively.

It may be worth noting here that all four species show seasonal dimorphisms. In *L. moluccellifolium*, *L. hybridum*, and *L. amplexicaule* this is probably a modification conditioned by the chances of germination. Some individuals germinate in autumn, over-winter, and flower in early spring, while others germinate in spring or summer and set seed and complete their life-cycle in the one season. The morphological differences between the seasonal forms are most marked in *L. moluccellifolium*. In *L. purpureum* there are biotypes with quite different periodicities. These differences are genotypically determined as shown both by cultural and hybridization experiments. One biotype is an obligatorily winter annual and in this there is only one generation a year. Other biotypes are summer annual or facultatively winter annual and generally develop several generations a year. The winter annual types over-winter as plants in the vegetative condition, the summer annuals, as a rule, over-winter as seeds.

Salicaceae

SALIX spp. (willows). The willows form a group which shows a great deal of variation from specific norms. There is now little doubt that, so far as British material is concerned, this variation is mainly due to hybridization between a limited number of morphologically well distinguished species. The cytogenetical facts are interesting because they show that the position in willows is different from that in the roses since, with only minor exceptions, there are no cytological abnormalities. It is not, however, entirely clear why, with hybridization frequent, the species as a whole have kept so distinct. Quite possibly the species are of old standing and have been well sifted by natural selection to suit special environments. Owing to man's activities these environments are less distinct than they were under natural conditions and the creation of numerous but scattered environments of intermediate or synthetic type tends to bring species in such juxtaposition that crossing between them now occurs much more readily than it did in the past. This is merely a suggestion made after recent visits to flooded quarries and other excavations filling with willow scrub !

There are two basic numbers of chromosomes among British willows : $x=19$ and $x=22$. Diploid, tetraploid, hexaploid, and other polyploid numbers have been recorded in the genus. The following tabulation shows the range in chromosome numbers of species :

$x=19$		$x=22$
Diploid $2n=38$	Tetraploid $2n=76$	Diploid $2n=44$
S. triandra	*S. pentandra*	*S. triandra*
S. purpurea	*S. fragilis*	Tetraploid $2n=88$
S. viminalis	*S. alba*	*S. phylicifolia*
S. capraea	*S. aurita*	*S. triandra*
S. repens	*S. atrocinerea*	Octoploid $2n=176$
S. lanata	*S. lapponum*	*S. glauca*
S. herbacea	*S. capraea*	
S. lapponum	Hexaploid $2n=114$	
S. aurita	*S. andersoniana*	
S. myrsinites	*S. phylicifolia*	
Triploid $2n=57$	Decaploid $2n=190$	
S. daphnoides	*S. myrsinites*	

This table shows that polyploidy has occurred in both series. *S. alba* and *S. fragilis* are said to be allotetraploids, i.e. they originated by doubling of the chromosomes following a cross. There are examples of polyploidy within one species in *S. capraea*, *S. lapponum*, *S. aurita*, and *S. myrsinites*. In *S. triandra* variants occur with $n=19$, $n=22$, and $n=44$.

The chromosome numbers in hybrids are interesting, as the table on page 232 shows—the figures in brackets giving the parental contributions to the total numbers.

Where the parents have the same chromosome number, and the uniting gametes the same reduced number, behaviour on reduction in the hybrid can be quite normal. In other crosses it often shows irregularities.

The willows are dioecious, the male and female flowers being on different individuals, with rare exceptions. XY chromosomes (sex-

Hybrid	*Somatic chromosome number*
S. lapponum × *S. viminalis*	38 (19+19)
S. herbacea × *S. lapponum*	38 (19+19)
S. lapponum × *S. capraea*	38 (19+19)
S. lapponum × *S. aurita*	38 (19+19)
S. capraea × *S. viminalis*	41 (19+22)
S. aurita × *S. atrocinerea*	76 (38+38)
S. capraea × *S. atrocinerea* (F_1 and F_2)	76 (38+38)
S. purpurea × *S. viminalis*	{38 (19+19) 57 (38+19)
S. purpurea × *S. aurita*	38 (19+19)
S. aurita × *S. phylicifolia*	63 (44+19)

chromosomes) have been reported in the males of *S. aurita*, *S. atrocinerea*, and *S. andersoniana*. Careful examination has shown that in diploid, tetraploid, and hexaploid species alike only one pair of sex-chromosomes is present.

Some very striking results have been obtained in Sweden in the building-up of hybrid willows under fully controlled pollinations. Most of the species used occur also in Britain and a few details concerning one of these synthesized hybrids are worth giving here. Eight different species were hybridized together according to the following plan :

1. *S. purpurea* × *S. daphnoides*
2. *S. repens* × *S. aurita*
3. *S. phylicifolia* × *S. nigricans*
4. *S. viminalis* × *capraea*

These separate crosses were successfully made. Then two more successful crosses were made :

5. *S. purpurea-daphnoides* × *S. repens-aurita*
6. *S. phylicifolia-nigricans* × *S. viminalis-capraea*

A final cross was :

S. purpurea-daphnoides-repens-aurita × *S. phylicifolia-nigricans-viminalis-capraea*

Two of the six shrubs of this " eight-species " willow were examined cytologically and found to be tetraploid (2n = 76) and to have very regular reduction divisions. The genes of one-third of the twenty-four species of *Salix* occurring in Sweden were united in one plant which was named *S. polygena*! The female shrubs are extremely fertile with pollen from male shrubs and 200 offspring of the original *S. polygena* have been raised, but the plants have grown very slowly and many are weakly developed.

Cyperaceae

CAREX spp. (sedges). Reference is made to this genus because of the irregular series of chromosome numbers recorded for different species. Thus, it has been found that in various species, $n = 9$, $n =$ every number from 14 to 42 inclusive, $n = 55$, and $n = 56$. The numbers cannot be arranged in a series of multiples such as would indicate polyploidy, or the multiplication of whole sets of chromosomes. Great differences in size are often found between chromosomes in the same nucleus. Groups of adjacent chromosome numbers are obtained when the species are arranged according to accepted systematic relationships. One hypothesis is that, in general, the higher numbers have arisen through duplication of entire chromosomes. In species formation in *Carex*, one of the most important processes has on this view been the formation of mutants with one extra chromosome which, in a later generation, have given rise to mutants with the diploid number increased by two. Another hypothesis is that there has been polyploidy followed by loss or, more rarely, addition, of individual chromosomes.

Gramineae

LOLIUM PERENNE (rye-grass) and allied species. A great deal of genetical and cytological work has been done on grasses because of their economic importance as grain- and fodder-producing plants. An outline of results obtained with rye-grass will indicate the kind of investigations which have been undertaken. *Lolium perenne*, as its botanical name indicates, is a perennial. It is usually regarded as the most desirable of all grasses in pastures in this country. There are a very large number of variants differing in such characters as habit of growth, rate of seasonal development, height, and time and degree of flower-stem production. Amongst characters of agricultural importance are : (*a*) those connected with behaviour in the seedling year, such as rapidity of growth and degree of tiller formation ; (*b*) date of emergence of the inflorescence in the first harvest year ; (*c*) number of inflorescences in the aftermath ; and (*d*) stemminess of cut herbage, most conveniently taken as the ratio of flowering tillers to non-flowering tillers ; and (*e*) nature of responses to grazing or cutting.

Rye-grass, like grasses in general, is naturally wind-pollinated. The anthers dehisce to shed the pollen after the stigmata are exposed. Many variants are self-sterile to their own pollen, and even in self-fertile stocks cross-pollination is very likely to occur under field conditions. Plants derived by cross-fertilization are superior in vegetative vigour to those resulting from self-fertilization. It has been shown that there are limits to cross-fertilization within a population of *L. perenne* due to cross-sterility or partial cross-fertility, although in compatible combinations such plants are sexually functional. In seasonal productivity it has been shown that cross-fertilization gives an increase over self-fertilization ranging from 37 to 224 per cent. Progeny from selfing often show marked lack of vigour, though such plants when intercrossed may give very vigorous F_1 families.

Several characters have been studied in *Lolium perenne*, with results that can be expressed factorially. Individual plants vary in intensity of green colour, but, with very rare exceptions, the lower part of the living sheath of the non-flowering tillers is of a distinct red or purplish-red colour. This is a very useful character in aiding determination of the species in the vegetative condition. Two plants were found which, on selfing, gave rise to both red base and non-red base seedlings. Breeding results indicated normal complementary factors. Two factors, C and R had both to be present to produce colour. Thus plants ccRR and CCrr were each non-red, while the double heterozygote CcRr had red-base tillers. A full green plant gave rise on selfing to three distinct types of seedlings : surviving green, non-surviving green, and yellow-tipped albinos. Two pairs of factors were involved and were represented by the symbols LlYy. In another plant a different pair of lethal factors, designated L_1l_1, was found to be responsible for survival or non-survival of green seedlings. On intercrossing the two strains, non-surviving seedlings could have any of the following constitutions : llL_1L_1, LLl_1l_1, llL_1l_1, Lll_1l_1, or lll_1l_1.

Lolium perenne has been crossed with other species of the genus. Six species, all diploids with 2n = 14, are recognized in the genus, and of

Brian Perkins

GROUNDSEL, *Senecio vulgaris* (Compositae). Therophyte. Waste ground, Kew, Surrey

Brian Perkins

OAK, *Quercus robur*, and BEECH, *Fagus silvatica* (Fagaceae) showing winter buds of phanerophytes. Mickleham, Surrey

PLATE XX

Brian Perkins

Ground Ivy, *Glechoma hederacea* (Labiatae). Chamaephyte in winter condition. Mickleham, Surrey

the fifteen interspecific crosses theoretically possible eleven have been made artificially. The results of crossing *L. perenne* and *L. temulentum* may be taken as an example. F_1 hybrids were obtained from reciprocal crosses between these two species. The type of grain produced varied significantly according to the direction of the cross. Seed germination and plant establishment were very poor when *L. temulentum* was used as the female parent, but good when *L. perenne* was so used. The established F_1 hybrid plants from the reciprocal crosses were similar in all essential characters. They were functionally male-sterile but could be used as female parents in back-crosses with either parent species as male, but seed-setting was always low. Smoothness of stem, relatively short glume, and awnlessness, all *L. perenne* characters, are dominant or partially dominant over the corresponding *L. temulentum* characters. The leaf-folding (conduplicate vernation) of *L. perenne* is at least partially dominant in the early growth stage. The young plants more closely resemble *L. perenne* in growth habit in the early stages, but in some respects the mature plants more closely approach the general appearance of *L. temulentum*. Development is rapid in the hybrids but time of flowering bears a definite relationship to the *L. perenne* parents. In mode of flowering the hybrids resemble *L. temulentum*. They are capable of vegetative regeneration after reaching full maturity, and thus differ from *L. temulentum* but agree with *L. perenne*.

Lolium perenne will also cross with species of fescue, such as *Festuca pratensis* and *F. rubra*. The former hybrid is supposed to be *Festuca loliacea* and if so is frequently recorded in the wild. These intergeneric hybrids are not easily made artificially and are mainly or entirely sterile.

Postscript to the Chapter

This chapter, it may be complained, is " too full of meat." It has, however, been prepared intentionally as a help to the naturalist desirous of some guide as to what is being done in the detailed study of the cytogenetics of British plants. The examples have been chosen carefully to indicate how diverse are the results already achieved. More important is a realization of how wide is the field for future research and how much more must be done by concentrated investigation before we understand the composition of British plant life. It is only by such experimental and observational research that problems of the origin and relationships of our flora can be satisfactorily solved. The field naturalist can help by his studies on populations in the wild and by providing localized material (as plants or seeds) for experiment and laboratory investigation. In particular he can determine the normal structure and behaviour of species under British conditions and can draw attention to abnormalities. There is abundant material at our doorstep, for not one-twentieth of British flowering plants have been investigated in as much detail as have many of those referred to in this chapter.

CHAPTER 13

CONTINUING CHANGES IN BRITISH PLANT LIFE

THIS CHAPTER includes speculations based on evidence of what has happened in the past and is happening now.

That the interaction of similar causes will lead to similar results in the future as in the past is the working hypothesis of science. It is a logical corollary that, given full knowledge of the past, one could correctly foretell the future, assuming that no new factor intervened. Since our knowledge of past and present events and their sequences is very incomplete a correspondingly very large element of doubt remains in any estimate of what will happen in the future. Reasoned speculation has, however, three advantages. Firstly, it enables planning on a rational basis, and if such planning has a degree of elasticity correlated with the degree of our knowledge (and therefore also with the degree of our ignorance) it gives us what is for the time being a maximum possible control of events. Secondly, it makes us build what knowledge we have into a unified whole, which may be incomplete and rickety in many parts but is not chaos. Thirdly, it strikingly shows up the gaps in our knowledge and suggests what are most important lines for future research.

We must consider how these general principles apply to our studies of British plant life. Though for convenience of treatment we have analysed our subject into chapters it has been kept constantly in mind, at the risk of some repetition, that our flora and vegetation are as we find them because of constant interaction between the plants themselves and their environment. To say that one is more important than the other is, in any general sense, a misunderstanding of what is involved in "cause and effect." One cannot have an environment without something to environ. What a plant without an environment would be cannot be imagined, at least it would be something not botanically recognizable. We have always to keep in mind the dual aspect of plant life: the plants and their surroundings; "nature and nurture"; genetics and ecology. There are two sets of changing factors, and to ignore either must give untrue pictures of the past, present, and future of plant life. The scientific method of abstracting, of isolating problems for study, is justified by its results, but we must recognize its dangers. Synthesis must follow analysis. Let us, then, consider what we can learn from changes in environment and from changes in genetic constitution of plants, and then attempt to synthesize the two groups of facts in such a way that we obtain some idea of the most likely future changes in the plant life of the British Isles and how we may best control them.

The environment of plants is, as we noted in Chapter 8, conveniently considered under the headings climatic, edaphic, and biotic. Our climate has fluctuated since temperate plant life became established here after the Ice Age, but there is no conclusive evidence of any major change or trend of change in the last 10,000 years. In quite recent times there has been a prolongation of autumn, a retardation of the on-coming of winter frost and snow. If this be maintained, and particularly if it be correlated with late spring frosts, it will alter the average dates of phenological phenomena: vegetative growth and reproductive processes will be extended towards Christmas while early spring development will be retarded. Perhaps most striking will be longer autumn retention of leaves by deciduous trees and shrubs and the slower bursting of buds in the spring. Autumn germination will be encouraged, spring germination will be delayed. The climate would increasingly favour the selection of hardy winter annuals or biennials. The effect of vegetation on climate is a controversial subject but, apart from microclimates, it is probably very slight. It is doubtful if such large-scale afforestation as might be undertaken in Britain would influence general temperature or rainfall conditions.

Edaphic or soil factors are of special interest in Great Britain because over so much of the country the soils are not mature. The inland ice of the Ice Age removed what soils there were and either exposed the mother rocks or re-covered them with heterogeneous deposits. Thus there are many stages in soil formation available for study, the more so since we have, in England in particular, over a relatively small surface area a very wide range of rocks of diverse geological ages and very different composition. Soil changes are, apart from man's interference, slow processes due to the continued action of climate and weather, combined with the activities of living organisms and the chemical and physical results of the decay of their dead remains. Owing to the diversity of the rocks, the differences in climate in east and west and north and south, and the correlated differences in plant cover, soil changes occurring in Britain vary from one locality to another. Three major trends can be recognized. Firstly, there is a tendency towards general maturity of soils as expressed in more sharp-layering, i.e. of the formation of more and more definite " soil horizons." Secondly, in our British climate, there is a tendency towards a general increase of acidity, a lowering of the pH, in the superficial layers. Thirdly, over much of the country, particularly in the south, the soils are becoming drier, less water-logged. This last change is probably due, at least very largely, to man, and might therefore be included under the heading " biotic." Drainage of farmland or potential farmland has certainly been accentuated during the Second World War, as is well exemplified by the work of Italian prisoners on the Ray and Cherwell with the idea of draining Otmoor in Oxfordshire. However, even before 1940 a general lowering of the water-table was noticeable in the south of England. Whether entirely due to man or greatly aided by his preliminaries to cultivation, from the Middle Ages onwards, our marshes, fens, lakes, and ponds have been shrinking and sometimes disappearing, and with them have gone habitats for the aquatic and marsh flora and vegetation. Even preserved areas, such as Wicken Fen, must in time be affected by increased withdrawal of water from their surroundings. Biotic factors are the most complex of all environmental factors affecting plant life. We have seen how vegetation, by its own interaction with the soil and climate, naturally changes from a minimum to a maximum control of the habitat. Such succession may be primary or secondary and the vast majority of seres (individual examples of succession) in Britain are, and for the future must be, secondary. These facts have to be remembered in considering proposals for retaining areas as nature reserves. If such areas be left alone there will be changes towards a climax, unless this is already reached. It is somewhat debatable how far, in areas left entirely to themselves, non-native plants would be exterminated, whether in the climax some aliens would be found to have established themselves or whether all would have succumbed to competition with the native plant life. The probability is that the native dominants would re-assert themselves but that a number of non-natives would survive in the earlier stages of the succession and as subordinates in the climax.

The human factor is the most important biotic factor now acting on our plant life, and is the most incalculable of all. Man has introduced a large number of very miscellaneous species which have become established members of our flora. These reproduce themselves, and in many existing habitats compete successfully or on equal terms with native species. Indeed, on a strict interpretation of the adjective "native," our plant lists would be very much reduced if only species were included about whose native status there was no doubt at all. As we saw at the end of Chapter 5, from Neolithic times onwards man has introduced weeds with his crops and the increase of commerce through the ages has brought in seeds and other propagules from all parts of the world. A large proportion of the weeds of arable land and the denizens of waste places are man-introduced and man-naturalized colonists. This process goes on and will continue indefinitely. It is generally true that the less the vegetation is disturbed by man the fewer in numbers and in species are the artificially introduced plants. By the destruction of forests and the draining of marshes, the making of roads, canals, railways, and aerodromes, the periodical ploughing of land, the planting of crops, from cereals, sugar-beet, or potatoes to forest trees, and in many other ways, man modifies and will continue to modify the environment for plant life and for himself. Besides enriching our flora while destroying our primitive vegetation almost to vanishing point, we have reduced the local occurrence of many species —especially of woodland and water plants. It is interesting, if depressing, to examine lists of plants and floras published a century or less ago for areas now very nearly or quite swallowed up in our large cities. In 1869 it was stated that 58 species which had been recorded for Middlesex were then extinct in the county. This number has been added to since then. A list of 10 species has been published as having become extinct since 1597 in Great Britain. Some of these, however, possibly never were native species ; the two most important, *Senecio paludosus* and *S. palustris* were both East Anglian species exterminated in the drainage of the Fens. Thus, so far, there is not much to mourn under the heading of complete extermination of species within the British flora, but a larger number of species only just maintain themselves and without planned protection, rigorously enforced, will disappear in the near future.

Let us sum up the environmental changes likely to occur in the future and to affect our plant life. Climatic changes are slow, if they occur at all beyond fluctuations of somewhat colder or warmer and wetter or drier seasons or years. Soil changes are always progressive and naturally, assuming no unexpected secular change of climate, there will be a slow change to more acid soils, favouring the spread of such species as prefer or tolerate acid soils and the restriction in range of calcicoles (definitely lime-needing plants). The continued lowering of the water-table will favour plants of well-drained soils at the expense of damp-soil and aquatic species. The most drastic changes in our vegetation and flora will be due, as in the immediately past centuries, to man's destruction and construction. Plantations for forestry are replacing woods for game preservation. Softwoods (conifers) are replacing hardwoods as prominent features of the landscape. Meadows and pastures are ploughed to arable and temporary leys give hay and graze in place of permanent pastures. Grasslands are improved and attempts will one day be successful in staying and reversing the spread of bracken on our western and northern hills.

One last and most important possibility remains for brief discussion. There are signs that naturalists are becoming increasingly aware of the need to preserve their natural heritage, or what remains of it, and are pressing for legislation to this end. Can man curb or reverse his own modifying activities and "give nature a chance"? Quite obviously in these densely populated islands a large proportion of the surface has to be occupied by artificial constructions and, if the population increases or even remains approximately stable, the total of such completely modified areas will tend to increase. Mere rise in the standard of living, as this is interpreted nowadays, involves increased artificiality, whether for good or for ill. The only solution to his problems for which the naturalist can reasonably hope is that a sufficient number of suitable areas of large enough size shall be preserved from exploitation. Actually preservation raises more problems than are at first apparent and some of them require a good deal of research before they can be solved. To mention only one, and that in general terms: an area of herbaceous vegetation which it is desired to preserve as such would, if left with no attention, by the natural process of succession, be replaced by a climax woodland.

What are the possibilities of inherent changes, of evolution in the narrower sense, occurring in the British flora? On the scale of years three score and ten, or even of human history, natural evolution is generally very, very slow. With domesticated animals and cultivated plants man has hastened its speed but few if any of his products could survive open competition in the wild. Undoubtedly, too, the tempo of evolution is not always the same for every group of organisms and has varied much in the course of geological ages. For example, the Angiospermae (the flowering plants) perhaps originated in Jurassic or even Triassic times, but the evidence suggests they had a great evolutionary spurt during the Lower Cretaceous when they spread over and "conquered" the land surface of the globe. There is no indication anywhere in the world of any group of plants originating or evolving to dominate the Angiospermae. Within our own British flora we can trace only minor lines of evolution along which there has been change in the not too distant past. In some of the groups involved, we can suggest that changes are still continuing and can reasonably speculate on the future of their microevolution. It is convenient to consider our speculations under headings, according to the mechanism involved in producing the initial variations.

GENE CHANGES may be accepted as probably resulting in the most basic variations for long-range evolution that have so far been at all adequately studied. They are numerous, may affect any organ or any function, are heritable, can cause the appearance of a new structure or of new behaviour, or both (though their expression may remain hidden indefinitely), the resultant new characters may be large or small differences from what was there before, their effects may be cumulative, in short, they provide variants very suitable for natural selection to act upon either drastically or slowly. A good example of gene changes is shown by the variants of *Silene maritima* and *S. cucubalus*. In British material there is no indication of chromosome changes. Indeed, throughout the large genus in all its range, so far as species have been cytologically examined, polyploidy is extremely rare, and all the British species have uniformly the same chromosome number, $2n=24$. Yet we have seen (Chapter 12, pp. 196 *sqq.*) how very variable the two common bladder campions are. There is no reason to suppose that their variability has all been discovered or studied, even for British populations, or that it has yet reached its limits. A large proportion of variants must of necessity, if genetic variations (or mutations) occur in the main at random, be less well adapted to existing conditions than the normal structure or function which has been subjected to very long-continued selection, and in this sense fits into the general scheme of habit, behaviour, and life-history which makes the organism already well adapted. There is some evidence that populations of *Silene cucubalus* have changed their average constitution

Brian Perkins

BLADDER CAMPION, *Silene cucubalus* (Caryophyllaceae). Banstead Downs, Surrey
Hemicryptophyte in winter condition

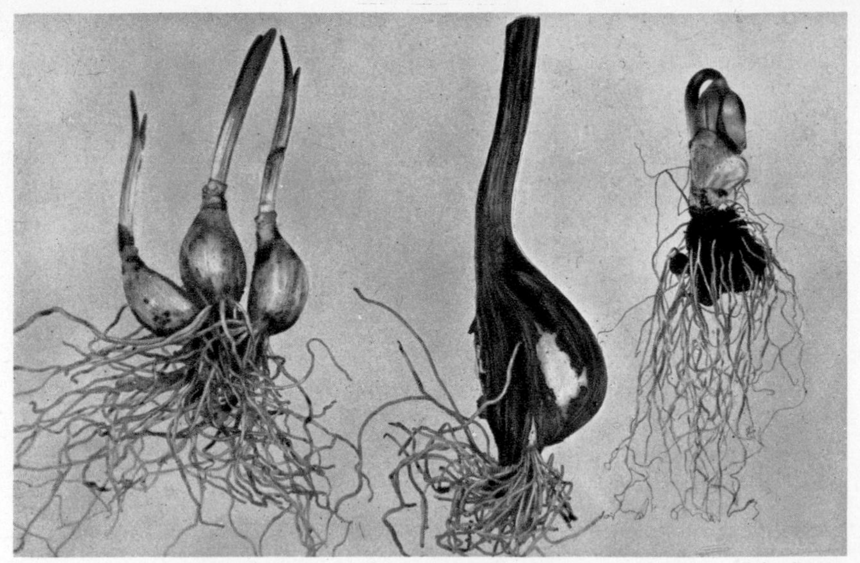

Brian Perkins

DAFFODIL, *Narcissus pseudo-narcissus* (Amaryllidaceae)
AUTUMN CROCUS, *Colchicum autumnale* (Liliaceae), and
WINTER ACONITE, *Eranthis hyemalis* (Ranunculaceae)
Material cultivated at Kew, showing winter condition of geophytes

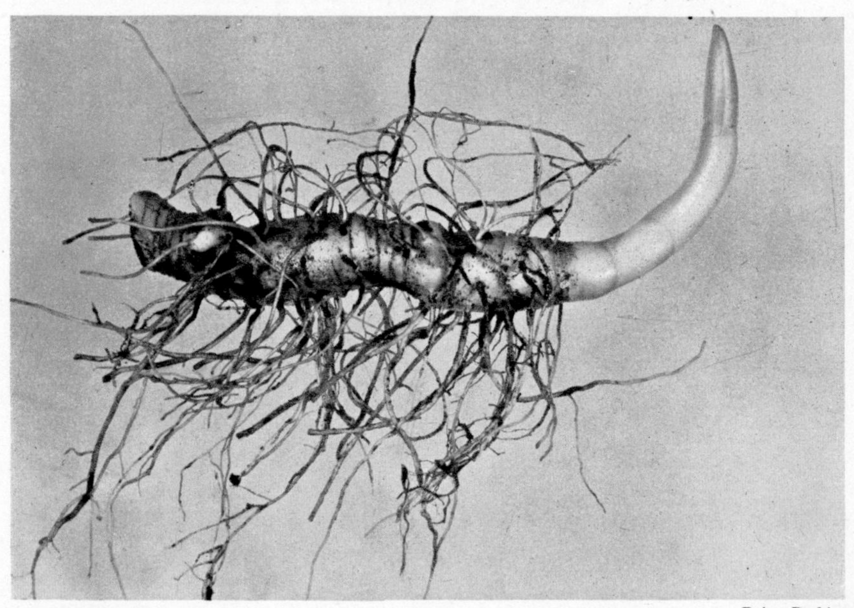

Brian Perkins

SOLOMON'S SEAL, *Polygonatum multiflorum* (Liliaceae). Norbury Park near Dorking, Surrey. Rhizomatous geophyte in winter condition

as the species has extended its range over Europe. In Britain it is still relatively rare towards the north. It will probably develop, by mutation and selection, physiological characters which will enable it to spread throughout the north and north-west.

Senecio vulgaris, *Ranunculus acris*, and *R. ficaria* are also species in which genetic variation has been studied. From observations on numerous wild populations and preliminary experiments with a number of other species, the generalized statements are made that every species is variable and that most are genetically much more variable than is known or realized. In particular, in the British Isles we have quite a number of species which here reach the northern limit of their ranges without occupying the whole country; frequently they are restricted in range to the south. Any one of these might so mutate as to be able to extend or change its range.

It has been suggested that the appearance of a long-homostyle mutation in populations of the common primrose (as in the Somerset populations referred to on p. 225) may change the character of the primrose in regard to heterostyly, perhaps over a wide area.

CHROMOSOME CHANGES may give more immediately spectacular results than gene changes. The rice-grass (*Spartina townsendii*) (Pl. 48, p. 223) originated, according to the available evidence, by polyploidy following hybridization on the south coast of England and has spread, and is still spreading, at a remarkable rate. *Spartina townsendii* was first recorded in 1878 from Southampton Water. Cytological investigation has shown that it is an allotetraploid which arose, with $2n = 126$, by the doubling of the chromosomes of a cross between *S. stricta* ($2n = 56$) and *S. alterniflora* ($2n = 70$). The two water-cresses (see p. 191) are morphologically and cytologically distinct and one, *Nasturtium uniseriatum*, appears to be an allotetraploid, as is also *Galeopsis tetrahit*. The two common valerians known from Britain represent tetraploid and octoploid races and have generally different distributions. One may expect further examples of auto- and allopolyploidy to be discovered in the British flora and polyploidy may suddenly start in genera (as *Silene*) or in species (as *Trifolium pratense*) where it is not yet known in Britain.

Another chromosome change is the addition or subtraction of single chromosomes. This appears to have happened in the course of the evolution of the sedges (*Carex* spp.) but is otherwise probably rarely of evolutionary importance in British genera, except that it may have been of greater importance in the early history of some of them and resulted in the foundation of now distinct chromosomal series.

SPECIAL REPRODUCTIVE MECHANISMS are exemplified by the dog-roses (see p. 210). The peculiar loss of chromosomes in the formation of male gametes and reduction in number in only a part of the chromosomes at formation of embryo-sacs allow of fertilization and reproduction in the offspring of the same number of somatic chromosomes as in the parent or parents. This gives a degree of cytological stability and of maternal inheritance but allows of diversity through hybridization. Such a specialized mechanism is rare and the chances are against its repetition. However, the diversification of British roses through such a mechanism is not exhausted and new varieties will continue to appear, some surviving and spreading locally or widely.

In the evening-primrose genus (*Oenothera*), only alien in the British flora, a different mechanism has evolved—one that has been termed an " evolutionary blind alley " because it allows of little, and decreasing, future variability. It is merely mentioned here to emphasize the possibility that other reproductive mechanisms, variants of the full normal sexual process, may yet be discovered or may yet arise in plants. It appears doubtful if any could have long-range success, given a continuation of existing environmental conditions. Combining our present knowledge of cytogenetics, palaeontology, and systematics, we can conclude that normal sexual reproduction has been the mechanism of evolutionary success, though there has been plenty of variation around the central motif.

HYBRIDIZATION is a term which has been used with both wider and narrower meanings. In the latter, it is restricted to sexual crossing by two organisms accepted as belonging to distinct species. In the former, the crossing of two different genotypes results in hybrids. These are extreme definitions and it is usually sufficient in practice to define the parents to the degree considered necessary and to let the " hybrid " or " non-hybrid " nature of the offspring take care of itself. However, crossing of different genotypes results in mixing of parental genes and may result in their segregation in later generations, if any.

Re-combinations are the more numerous and the more varied, the greater the differences between the parental genotypes. Hence, if sterility does not intervene, crossing between different species gives, in general, greater recombinational variability than crossing between varieties. We have spoken of the hybrid swarms in Britain of *Geum*

(p. 209), *Viola* (p. 194) and *Centaurea* (p. 221). Gene-exchange may occur and genes may " flow " through a population as a result of hybridization. Hybridization is restricted in its occurrence and in the permanency of its effects by isolating barriers and by natural selection. Plants must be within pollinating distance of one another for cross-pollination to occur. There must be no incompatibility or other kind of sterility between them for cross-fertilization to be successful. Offspring (hybrids, heterozygotes, or whatever one calls them) must be able to establish themselves to maturity, and must be fertile, if crossing is to have any long-range influence in evolution. All these conditions are least likely to be fulfilled under natural conditions of climax vegetation. Hence the importance of catastrophes such as the Ice Age or man.

APOMIXIS occurs in all or a large proportion of British hawkweeds (*Hieracium*), dandelions (*Taraxacum*) (Pl. 47, p. 222), and lady's-mantles (*Alchemilla*). It is frequent in the blackberries (*Rubus*) and occurs in the lesser celandine (*Ranunculus ficaria*). It is very likely of casual occurrence in many more species than is at present known. It functions in somewhat less than 1 per cent of the ovules in some stocks of common buttercup (*Ranunculus acris*). That apomictic blackberries can show segregation suggests that the disadvantage of apomixis in not allowing variability in the offspring is not absolute. It is difficult to believe that variation has entirely ceased in the apomictic British dandelions and one suspects either that sexually reproducing *Taraxaca* remain for discovery in our British flora or that some mechanism of variation awaits investigation. Some of our species of southern origin and southern range belonging to other genera may develop or increase the apomictic development of seeds and this may enable their ranges northwards to be extended.

CYTOPLASMIC CHANGES may prove more important than hitherto generally recognized. We have seen (p. 217) how the cytoplasm plays a large part in the appearance of characters in the offspring of certain crosses in *Epilobium*. Femaleness or male sterility is in some plants dependent, at least in part, on the cytoplasm. It is possible that a good many British plants (particularly members of the Labiatae) are showing an increasing tendency to gynodioecism—at least this is a subject requiring much more research both by statistical studies on wild populations and by controlled experiments. A further possibility is that genes, if so we term them, outside the nucleus—the terms " plastogenes " and "plasmagenes" have been suggested for them—may be more easily changed by outside environmental factors than genes protected in the chromosomes.

Variation may affect any structure and any function. Mutation, it is often stated, is random. The thoughtful naturalist, however, sometimes wonders whether there is not some agency or mechanism, besides natural selection, which in the long run directs evolution along some limited lines. It is possible that the chemical and physical constitution of the chromosomes and cytoplasm determine that certain mutational changes shall occur in such sequence that characters are produced in an orthogenetic (straight line) series. The subject is too theoretical and technical to discuss further here, but it is as well not altogether to stifle such heterodox doubts of accepted theories as sometimes intrude when contemplating past changes and the possibilities of future changes in British plants.

We have considered in outline two more or less frequently changing sets of variables—that of the external environment and that of the inherent and inherited constitution of any given plant or plant kind. In producing plant life, whether in the development of the individual from fertilized egg (zygote) to maturity or in the course of evolution of the species or other taxonomic unit, the external environment and the genetic constitution are always interacting; changes in either affect the result, changes in both may be even more significant. Let us, in concluding this chapter, consider how possible or probable environmental changes may give recombinations or mutations from what is now the normal a chance to spread and to change the make-up of species and of vegetation.

Man is the most drastic in action of environmental factors for plant life in the British Isles. Intentionally or unintentionally he introduces all sorts of new species into our flora. By his activities he changes habitat conditions on a wholesale scale. Man breaks down the barriers resulting in natural geographical isolation; he disrupts the barriers of natural ecological isolation. One result is that species hitherto distinct are brought together and sometimes hybridize. This is happening in the genus *Centaurea* and quite probably in *Salix*, *Thymus*, *Trifolium*, *Bromus*, and *Poa*. One may speculate that some of the recombinations of characters will be better adapted to the habitats opened up or created by man than are any of the pure species.

Apart from hybridization, mutations also bring about the formation of new biotypes, some of which prove better adapted to man-made environments than do the parental biotypes. This has certainly happened with weeds, though the subject has received less study in this country than on the Continent. The fool's parsley (*Aethusa cynapium*), spurrey (*Spergula arvensis*), chickweed (*Stellaria media*), charlock (*Sinapis arvensis*), knotweed (*Polygonum aviculare*, in the broad sense), couch-grass (*Agropyron repens*), and vetches (*Vicia* spp.), it is suggested, would well repay study from this standpoint. Ruderals, plants of waste places and waysides, also need detailed and comparative study. Many of them are either newly introduced aliens or native plants which have invaded, and tolerate, new kinds of habitats. They may well have formed or be forming new ecotypes. The phenomenal spread of certain species in the last few decades—*Matricaria matricarioides*, *Senecio squalidus*, *Crepis taraxacifolia* are examples—suggests that some genetical changes may have occurred in these species. Very incomplete observations on the Oxford ragwort (*Senecio squalidus*) indicate that its variability is greater at Oxford than it is in at least some London districts.

Man is constantly changing his mode of life. Agricultural practice changes: there are new machines, new crops, new fertilizers, new chemical, mechanical, and cultural treatments for diseases, insect pests and weeds, and so on. On the Chalk and oolitic limestones, until some twenty or thirty years ago, the arable crops were often yellow with charlock (*Sinapis arvensis*). Spraying with sulphuric acid now restricts this weed to field margins—indeed it may be locally disappearing—but has allowed the spread of wild oats (*Avena* spp.), of which there is one species of recent introduction (*Avena ludoviciana*) which is tending to become a menace. If, like the bromes of the *Serrafalcus* group of the genus *Bromus*, it starts to vary and to hybridize with other species it may come to occupy a wide range of man-made habitats. A rather striking change of a somewhat different character has been kept under observation at intervals. Near Woodstock in Oxfordshire there is a network of green lanes, grass-grown tracks bounded by hedges, on oolitic limestone (mainly Cornbrash and Forest Marble). Forty to fifty years ago these lanes were grazed fairly hard by flocks of sheep which were usually in charge of a shepherd boy. The flora was rich with lime-needing plants—thyme, rockroses, quaker-grass, and the like. Now, probably owing largely to the increased cost of labour and the non-employment of boys, the lanes are not grazed. The vegetation has changed greatly and large areas are covered in summer with the tall growth of the erect brome (*Bromus erectus*). This species is very palatable to sheep but cannot stand grazing. In the absence of sheep it has dominated the vegetation and the plants show a wide range of variation. So far no natural selection of ecotypes has been observed, but genetically variable material is there in plenty and some combinations of characters may oust others or may extend the ecological range of the species.

CHAPTER 14

REVIEW, CONCLUSIONS, AND THE FUTURE

OUR STUDIES have covered a wide field, though almost limited to British plant life. The attempt has been made deliberately to introduce as many diversified aspects as space would allow with the object of showing how varied are the opportunities open to the naturalist for making further contributions to our knowledge of the composition, behaviour, and history of the British flora and vegetation. There is, however, a unifying plan running through the chapters in the desire to consider plant life, whether as flora or vegetation, as related to its environment and altering in correlation with this and its own internal potential of change. Systematically our flora is relatively well known at the species level. Much less is known regarding variation of species. Indeed, the published lists of varieties contain a hotchpotch of variants characterized by differences with a genetic basis and mere forms due to growth in different habitats. What is needed is, by extensive and prolonged observation and intensive experiment, to " get within the species "—to understand its genetic make-up, to learn its ecological reactions, and to appreciate its behaviour as a living changing population. It is only in this way that a fund of knowledge can be accumulated sufficient to solve what are commonly termed the " problems of species."

The grouping of plants into classes, mainly on the basis of their structure, the arrangement of classes in a logical hierarchy, and the giving of Latin names to classes defined by description of characters are termed systematics or taxonomy, and are essential because of the very large number of kinds of organisms. On the whole, the groupings worked out by taxonomists are highly satisfactory for a great variety of purposes, but with new knowledge they are continually capable of improvement. The " species " is usually regarded as the most important of taxonomic units. When the name of a plant is requested, it is usually the species name that satisfies the request. Like all taxonomic units the species has two aspects—a concrete and a theoretical. A given species is a population of individuals, whatever the doubts as to its boundaries, as to which individuals should be included and which excluded. If familiar with the flora of an area, we can usually place a very high proportion of the plants we find on any excursion definitely into one or another species as recognized by most taxonomists. This is a very good pragmatic test for the existence of distinguishable " kinds " in a flora, and many such are recognized as distinct by non-botanists who are in no way influenced by taxonomic tradition. On the other hand, the choice of criteria by which we decide what groups shall be called "species," "subspecies," "varieties," and so on, is very much a matter of convenience, even if it be scientific convenience. Needs, experience, accepted principles, information available, and subjective bias all play a part. In the present state of knowledge a certain flexibility of taxonomic treatment may be advisable and there is some tendency to recognize both " aggregate species " and " microspecies." " Aggregate species " can be divided into smaller units but these are not distinguished by constant combinations of well-marked characters. "Microspecies" are distinguished by few characters one from another or from some larger species. There is this one certainty : the more we can learn about the structure, behaviour, and life-history of the members of a group, however we designate it, the nearer we shall get to a generally satisfactory classification not only of the group in question but of all groups.

Very early in any research involving organisms, some scheme of classification is essential. Such a scheme may be very imperfect and only research itself can improve it. The ideal of a general classification is for classes to be based on as wide a range of characters as possible and to be of maximum use for all purposes in which the classes are concerned. In botanical classification in the past, external structural features have been used, mainly or solely, for vascular plants. Such a classification is basically not ideal. A preliminary classification has been termed an " alpha " classification in contrast to the perfect ideal of an " omega " classification. This latter would be based on all possible characters and all relevant data and is an ideal which can be approached but is unlikely ever to be fully attained. Many of the lines of research illustrated in previous chapters are helping to move plant classification a little way from " alpha " towards " omega " ; no more is claimed. It must be added that special classifications based on one or a few kinds of characters may be of very great importance for special purposes, including improvement of a general classification. Further,

PLATE XXIII

John Markham

DOG VIOLET, *Viola riviniana* (Violaceae). Near Northaw, Hertfordshire. May 1946

Eric Hosking

PRIMROSES, *Primula vulgaris* (Primulaceae); pin-eyed. Near Woodbridge, Suffolk April 1935

John Markham

PRIMROSES, *Primula vulgaris* (Primulaceae); thrum-eyed Northaw Woods, Hertfordshire. April 1938

since a general classification and its associated nomenclature have to serve a wide range of theoretical and applied purposes—as those of ecologists, cytogeneticists, physiologists, foresters, horticulturists, etc.—changes in it must be well-founded before they are made. In other words, taxonomic research must be progressive, yet in altering classification taxonomists must be reasonably conservative.

We have seen how valuable ecological and cytogenetical investigations are in throwing light on the constitution and limits of groups, on their origin, and on their history. There is, however, a very great deal of research needed before what has been called "speciation" (the origins of species) can be adequately discussed with sufficient well-studied examples for general laws to be traced and the facts used as a basis for a widely improved classification. Within our British flora hundreds of species await observation and experiment on the scale with which the bladder campions are being investigated. The common daisy, the bluebell, the yarrow, the wild thyme (or thymes), and many others are much more variable than is usually realized. Any one such species or group would repay intensive study from the combined standpoints of taxonomy, ecology, genetics, and cytology. In undertaking such research it is essential to have a well-considered plan for recording all observations, all experiments, and all results (negative as well as positive). The desirability of team-work should be considered. The Systematics Association has a Research Committee which is very willing to give advice and help to all serious students wishing to undertake investigations along the lines indicated above (see p. 263). Field naturalists engaged in more extensive studies can do much towards helping intensive research by recording observations and collecting seeds or other propagules and communicating them either directly to specialists or to a central research institution. Mutations may be particularly valuable, and somatic (body) mutations of British wild plants are particularly needed in the living state.

It is suggested that the detailed comparative study of a few species within one genus is likely to yield very interesting results. Presumably such species are more or less closely related genetically but they may be ecologically diverse. This is probably one of the best ways to investigate speciation. How far does specific diversification result in optimum exploitation of environments? What is the relationship between ecotypes and accepted specific units? A selection of small groups of species for observational and experimental analysis might well be made from such genera as: *Thalictrum*, *Papaver*, *Fumaria*, *Barbarea*, *Polygala*, *Cerastium*, *Hypericum*, *Geranium*, *Ononis*, *Vicia*, *Agrimonia*, *Pimpinella*, *Senecio*, *Campanula*, *Symphytum*, *Myosotis*, *Verbascum*, *Linaria*, *Scrophularia*, *Veronica*, *Salvia*, *Chenopodium*, *Polygonum*, *Rumex*, *Euphorbia*, *Allium*, *Juncus*, *Carex*, *Alopecurus*, *Calamagrostis*, *Deschampsia*, *Holcus*, *Bromus*, and *Brachypodium*. Pairs of species are well worth comparative study: *Bidens cernua*—*B. tripartita* and *Triglochin maritimum*—*T. palustre* may be suggested for investigation.

Again and again we have emphasized the dynamic aspects of British plant life. The flora and the vegetation are always changing. There are fluctuations from scarcity to abundance and from abundance to scarcity; there are long-range changes, species become rare or die out and others appear, some to become common; vegetation shows succession with varied details. It is only by long-continued observations and experiments and, most important, careful recording that we can learn the results, the courses, and the causes of such changes. How and when do aliens come in and what are the underlying causes for the successful establishment and spread of some of them? How are plants maintained with restricted ranges? *Stachys germanica* in Oxfordshire, the oft-quoted plants of Teesdale, and others, are examples of our ignorance on this problem. What are the causes of death of plants in the wild? We have very little information on this and practically none on a statistical basis. Even our knowledge of the death of seedlings is, with a few exceptions, exceedingly meagre. Till we know more about the causes of death many statements regarding natural selection are tentative or require limiting qualifications. How often is survival merely chance and how often the result of better adaptation?

One aspect of British plant life must be the concern of all naturalists—its preservation. We have seen that man has modified and continues to modify our flora and vegetation. Over large areas of our country this is unavoidable. Nevertheless a number of measures can be taken to prevent irreparable loss in flora and vegetation and in associated animal life. Such measures range from the institution of Nature Reserves and National Parks and the ownership of areas by the National Trust to smaller reserves and even local and individual prevention of destruction.

The conservation of our wild life has been very fully considered recently by several authoritative bodies who have published comprehensive (and exceedingly readable) reports. Three main objects of nature preservation have been distinguished : the aesthetic, the scientific, and the educational. Maintenance of the beauty and interest of characteristic British scenery has rightly a wide appeal. The scientific value of the study of animals and plants in the wild and in communities as natural and primitively natural as possible is shown by much of the contents of previous chapters. The educational values of *in situ* nature-study do not cease at school-leaving age. Readers of this book will not need to have them tabulated, but naturalists owe it to themselves, to their subject, and to their fellow citizens to instruct all who will listen or read in the pleasures, the mental and physical advantages, and even the utilities of an understanding of our wild life and hence the need of preserving and, in a reasoned and balanced manner, increasing the value of our heritage of woodland and moor, of dunes and pastures, of fens and mountains.

In considering the upkeep of Nature Reserves, attention must be given to the precise aims of preservation. If it be desired to maintain flora and vegetation in an existing or any given condition other than a natural climatic climax, there must of necessity be periodic interference, otherwise the plant life will change according to the laws of succession. Such changes, of course, also involve the animal life. It is reasonably proposed that, within a reserve, a part of the area shall be allowed to remain at or attain to a climax condition and the remainder be kept in such earlier seral stages as are deemed desirable. This in particular will mean the periodic cutting of brushwood, whether it be a seral scrub stage or undergrowth (for example, coppice) to high woodland. Ponds and rivers or streams may have to be dredged and herbaceous communities mown at certain intervals. This procedure is particularly necessary for " species reserves " when a main aim is the preservation of one or more particular kinds of plants (or animals). They can be preserved only by preserving the habitat range they tolerate.

MAPS SHOWING THE RANGE OF CERTAIN PLANTS IN THE BRITISH ISLES

THE FOLLOWING maps are intended to give a general idea of the vice-county range of certain of our native seed-bearing plants. They do not give details of distribution and further research may lead to their modification in some respects. The examples have been chosen to illustrate the main kinds of ranges as these are generally accepted in British floras. Definitions, additional examples, and a discussion will be found in the text (pp. 53 *sqq.*). Druce's *The Comital Flora of the British Isles*, Arbroath, 1932, has been used in the compilation of these maps.

1. Distribution of British Type: *Corylus avellana* (hazel)
2. Distribution of English Type: *Bryonia dioica* (white bryony)
3. Distribution of Germanic Subtype: *Frankenia laevis* (sea-heath)
4. Distribution of Atlantic Subtype: *Umbilicus pendulinus* (pennywort)
5. Distribution of Scottish Type: *Trollius europaeus* (globe-flower)
6. Distribution of Highland Subtype: *Loiseleuria procumbens* (trailing azalea)
7. Distribution of Intermediate Type: *Actaea spicata* (baneberry)
8. Distribution of Local Type: *Dianthus caesius* (cheddar pink)

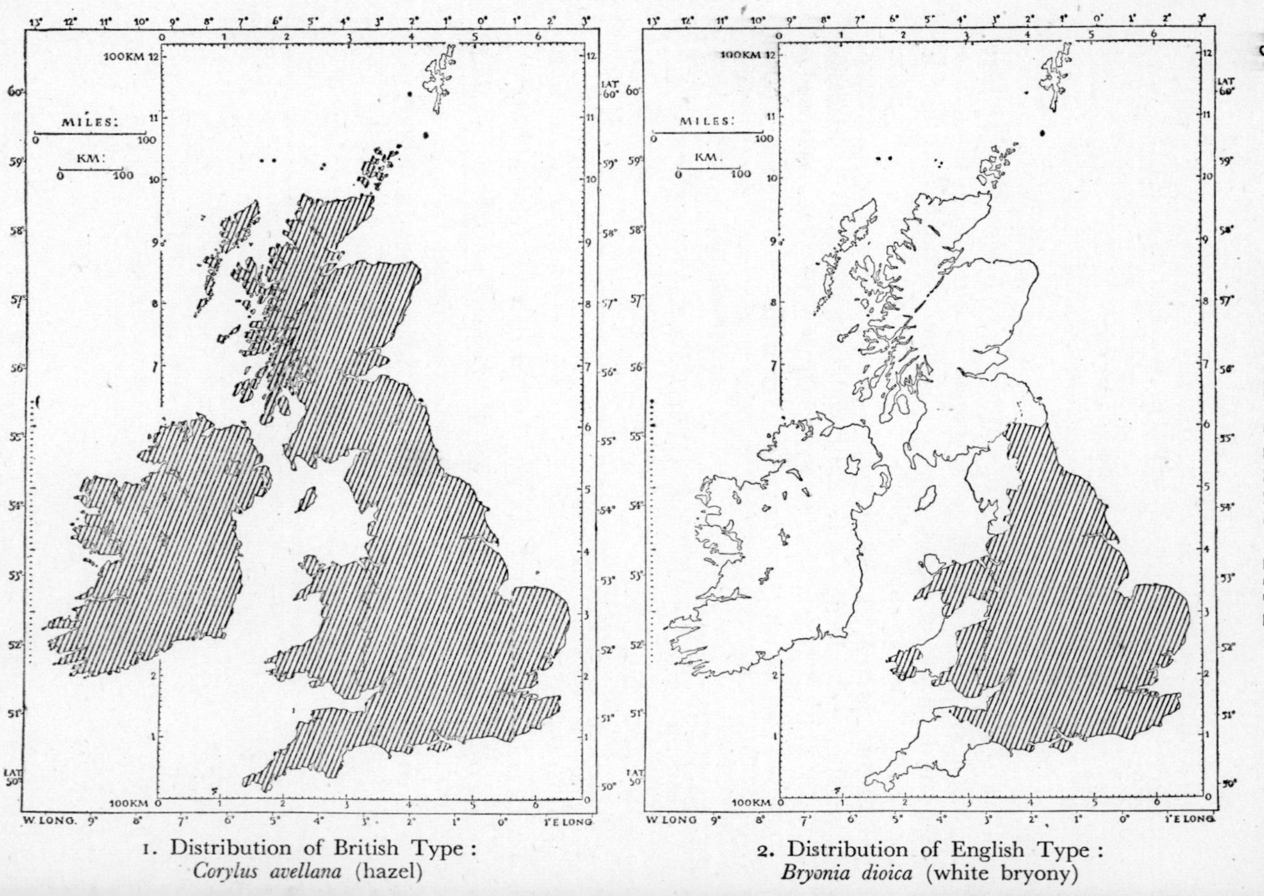

1. Distribution of British Type: *Corylus avellana* (hazel)

2. Distribution of English Type: *Bryonia dioica* (white bryony)

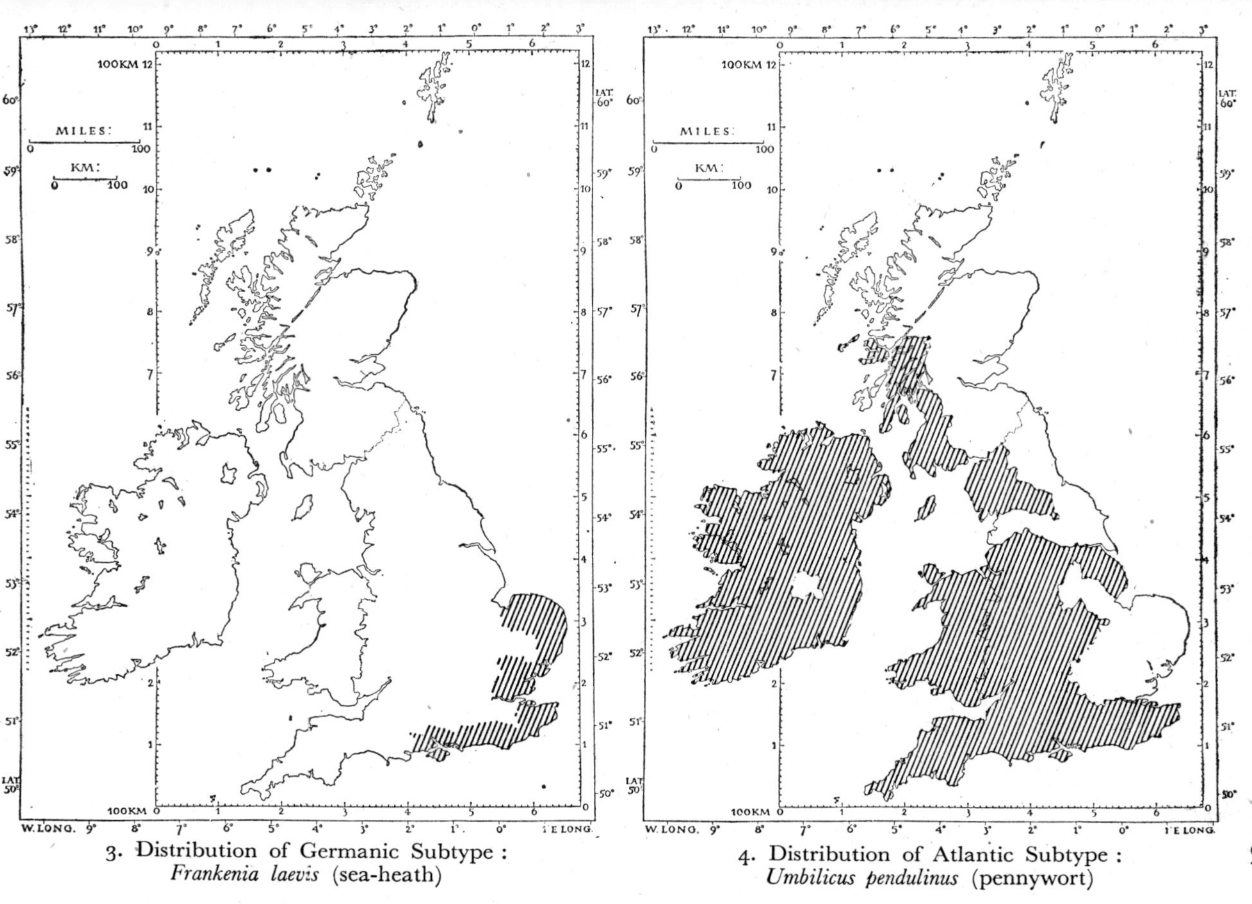

3. Distribution of Germanic Subtype: *Frankenia laevis* (sea-heath)

4. Distribution of Atlantic Subtype: *Umbilicus pendulinus* (pennywort)

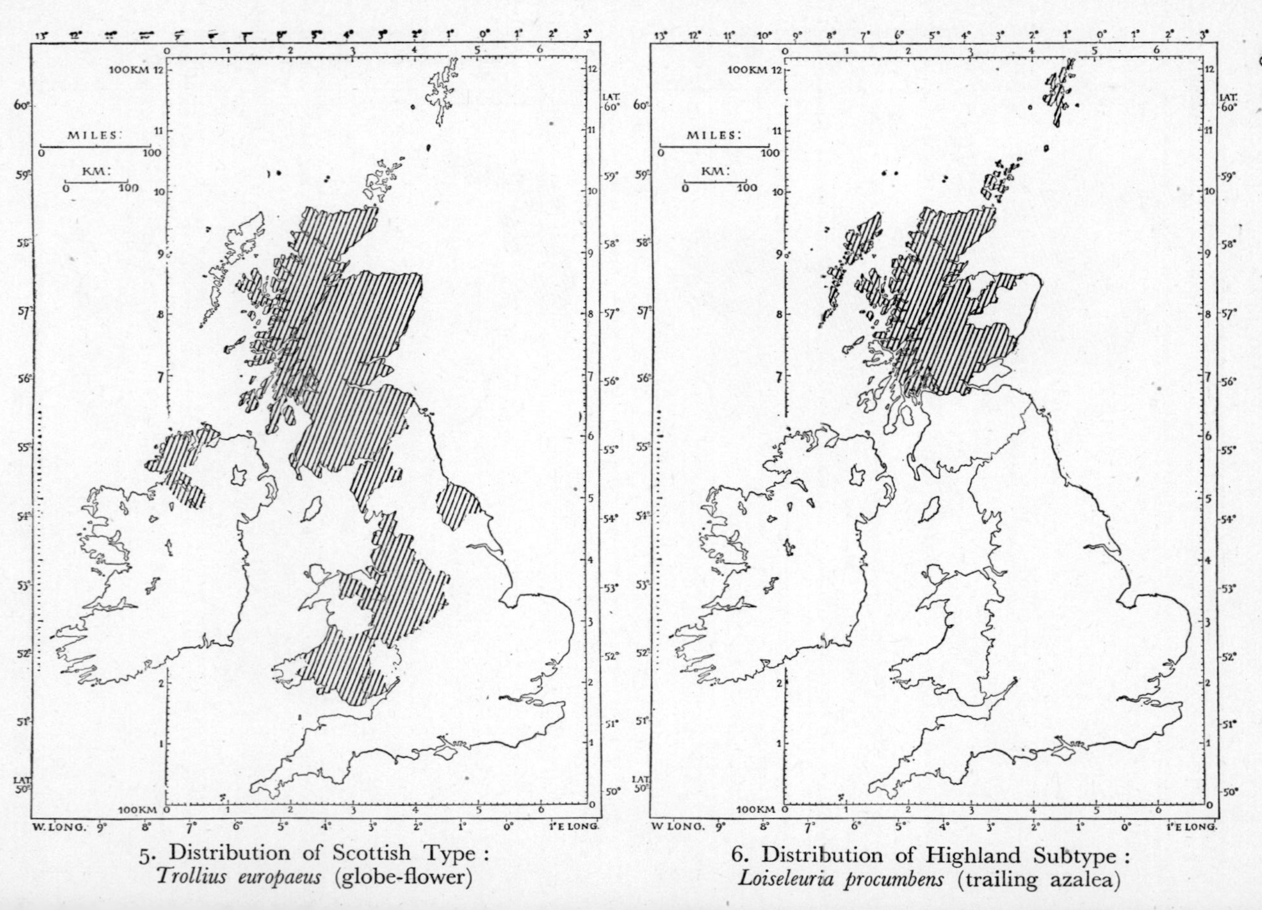

5. Distribution of Scottish Type:
Trollius europaeus (globe-flower)

6. Distribution of Highland Subtype:
Loiseleuria procumbens (trailing azalea)

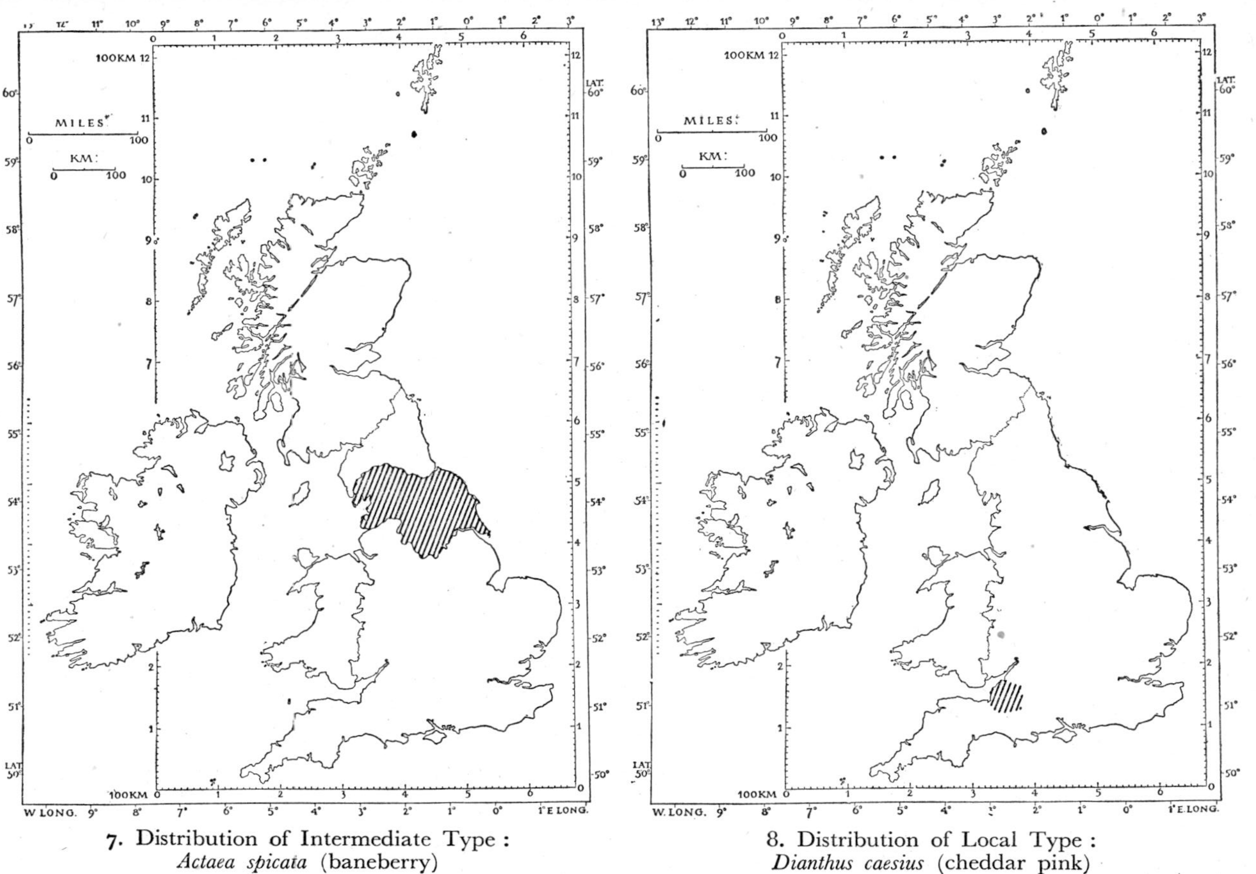

7. Distribution of Intermediate Type:
Actaea spicata (baneberry)

8. Distribution of Local Type:
Dianthus caesius (cheddar pink)

APPENDIX A

LIST OF VICE-COUNTIES

ENGLAND AND WALES

PENINSULA

1 West Cornwall with Scilly
2 East Cornwall
3 South Devon
4 North Devon
5 South Somerset
6 North Somerset

CHANNEL

7 North Wilts
8 South Wilts
9 Dorset
10 Isle of Wight
11 Hants, South
12 Hants, North
13 West Sussex
14 East Sussex

THAMES

15 East Kent
16 West Kent
17 Surrey
18 South Essex
19 North Essex
20 Herts
21 Middlesex
22 Berks
23 Oxford
24 Bucks

ANGLIA

25 East Suffolk
26 West Suffolk
27 East Norfolk
28 West Norfolk
20 Cambridge
30 Bedford and detached part of Hunts
31 Hunts
32 Northampton

SEVERN

33 East Gloucester
34 West Gloucester
35 Monmouth
36 Hereford
37 Worcester
38 Warwick
39 Stafford and Dudley
40 Shropshire

SOUTH WALES

41 Glamorgan
42 Brecon
43 Radnor
44 Carmarthen
45 Pembroke
46 Cardigan

NORTH WALES

47 Montgomery
48 Merioneth
49 Carnarvon
50 Denbigh and parts of Flint
51 Flint
52 Anglesey

TRENT

53 South Lincoln
54 North Lincoln
55 Leicester with Rutland
56 Nottingham
57 Derby

MERSEY

58 Cheshire
59 South Lancashire
60 Mid Lancashire

HUMBER

61 South-east York
62 North-east York
63 South-west York
64 Mid-west York
65 North-west York

TYNE

66 Durham
67 Northumberland, South
68 Cheviotland, or Northumberland, North

LAKES

69 Westmorland with North Lancashire
70 Cumberland
71 Isle of Man

SCOTLAND

W. LOWLANDS

72 Dumfries
73 Kirkcudbright
74 Wigtown
75 Ayr
76 Renfrew
77 Lanark and E. Dumbarton

E. LOWLANDS

78 Peebles
79 Selkirk
80 Roxburgh
81 Berwick
82 East Lothian
83 Midlothian
84 West Lothian

E. HIGHLANDS

85 Fife with Kinross
86 Stirling
87 South Perth with Clackmannan, and parts of Stirling
88 Mid Perth
89 North Perth
90 Angus or Forfar
91 Kincardine
92 South Aberdeen
93 North Aberdeen
94 Banff
95 Moray or Elgin
96 Easterness (East Inverness with Nairn)

W. HIGHLANDS

97 Westerness (West Inverness with North Argyll)
98 Argyll (Main)
99 Dumbarton (West)
100 Clyde Isles
101 Cantire
102 South Ebudes (Islay, etc.) and Scarba
103 Mid Ebudes (Mull, etc.)
104 North Ebudes (Skye, etc.)

N. HIGHLANDS

105 West Ross
106 East Ross
107 East Sutherland
108 West Sutherland
109 Caithness

NORTH ISLES

110 Outer Hebrides
111 Orkney
112 Shetland

IRELAND

113 (1) South Kerry
114 (2) North Kerry
115 (3) West Cork
116 (4) Mid Cork
117 (5) East Cork
118 (6) Waterford
119 (7) South Tipperary
120 (8) Limerick
121 (9) Clare with Aran Isles
122 (10) North Tipperary
123 (11) Kilkenny
124 (12) Wexford
125 (13) Carlow
126 (14) Leix
127 (15) South-east Galway
128 (16) West Galway
129 (17) North-east Galway
130 (18) Offaly
131 (19) Kildare
132 (20) Wicklow
133 (21) Dublin
134 (22) Meath
135 (23) Westmeath
136 (24) Longford
137 (25) Roscommon
138 (26) East Mayo
139 (27) West Mayo
140 (28) Sligo
141 (29) Leitrim
142 (30) Cavan
143 (31) Louth
144 (32) Monaghan
145 (33) Fermanagh
146 (34) East Donegal
147 (35) West Donegal
148 (36) Tyrone
149 (37) Armagh
150 (38) Down
151 (39) Antrim
152 (40) Derry

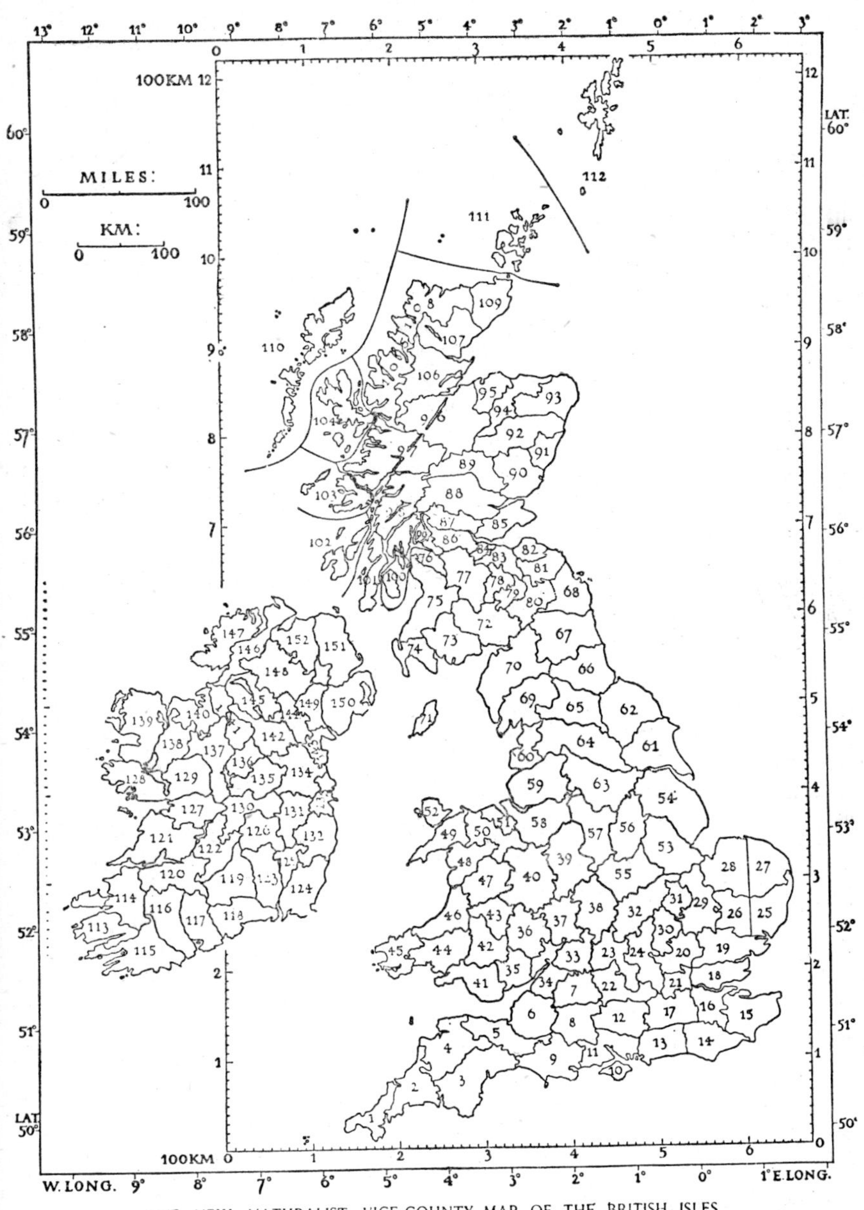

THE NEW NATURALIST VICE-COUNTY MAP OF THE BRITISH ISLES

APPENDIX B

TAXONOMIC METHODS

IN STUDIES of plant life it is essential to have determination as precise as possible of all plants investigated. Some groups require the aid of specialists but the naturalist should aim at learning to determine as many species as he can for himself. A hand lens, a dissecting microscope, and simple dissecting tools are essential. For such groups as the Algae, smaller Fungi, Lichenes, and Bryophyta a compound microscope and accessories are needed. Special instructions on how to work out the "taxonomic characters" are usually provided in text-books special to groups. Many students will find it of great value to form a herbarium, collection of microscope slides, or a series of drawings or paintings of the groups with which they work. The aim should be to make these complete for the purposes of the investigations in hand—they should be intensive and not extensive. It is worth stating that such collections may have permanent scientific value if properly made and properly preserved. It is, therefore, a matter of importance that when they have served the immediate purpose for which they were made they should be presented or definitely bequeathed to a scientific institution where they will be under good curatorship and yet available for use to future investigators.

Botanists wishing to become acquainted with others interested in the systematic study of British plants are advised to join one or more of several societies, of which the following arrange excursions and issue publications :

The Botanical Society of the British Isles. Address : The Secretary, British Museum (Nat. Hist.), S.W.7. For the study of vascular plants (seed-bearing plants, ferns, and fern-allies).

The British Mycological Society. Address : The Secretary, G. C. Ainsworth, B.Sc., Ph.D., F.L.S., London School of Hygiene and Tropical Medicine, Keppel Street (Gower Street), W.C.1. For the study of Fungi.

The British Bryological Society. Address : The Secretary, E. C. Wallace, 2 Strathearn Road, Sutton, Surrey. For the study of mosses and liverworts.

The Systematics Association caters for a wide range of interests in all matters connected with the relations between systematic botany and zoology and general biology, and in particular with evolution in the broadest sense. Information can be obtained from the Botanical Secretary, Dr. R. Melville, Royal Botanic Gardens, Kew, Surrey. In particular anyone wishing to undertake research work on British plant life should obtain help and advice from the Research Committee of the Systematics Association (Secretary, Dr. W. B. Turrill, The Herbarium, Royal Botanic Gardens, Kew, Surrey).

Since we have had in various parts of this work to refer to major plant groups or classes, a simplified classification of the plant kingdom is given below. This is not intended to be a critical tabulation of all the known kinds of plants but only a guide to the terms used for the larger divisions in most classifications of plants. Though the simpler groups are placed first and what are generally regarded as the more highly evolved follow in approximate sequence of probable evolutionary relationship, the scheme has no intention of representing the possible course of evolution of the plant kingdom, which must have been much more complicated than can be shown in a linear tabulation.

PLANT KINGDOM

SCHIZOPHYTA or "splitting plants," with no differentiated nuclei and no recorded normal sexuality :
- Bacteria.
- Myxophyceae (Cyanophyceae) or blue-green Algae.

THALLOPHYTA : plants with no differentiation into root, stem, and leaf; reproduction by spores of various kinds of which some are sexually produced :
- Algae—green, brown, and red, with chlorophyll.
- Lichenes—composite organisms of algal and fungal components.
- Fungi—including mushrooms and toadstools, and many other and often microscopic types, with no chlorophyll.

BRYOPHYTA : plants with two "generations" of which the spore-producing is attached to the sexual generation which has a green vegetative body often with leaves :
- Hepaticae or liverworts.
- Musci or mosses.

PTERIDOPHYTA or vascular cryptogams : plants with two quite separate generations, the spore-producing plant being the more conspicuous and differentiated into roots, stems, and leaves, and with wood and bast occurring as part of the anatomical structure :
- Lycopodiales or club-mosses.
- Equisetales or horsetails.
- Filicales or ferns.

SPERMATOPHYTA or seed-bearing plants, reproducing by seeds :
- Gymnospermae : with naked ovules (structures which, after fertilization, give rise to the seeds) and naked seeds, pollination being direct. Including the conifers (pines, larches, firs, spruces, cedars, etc.), cycads (not

British or European in existing flora), and the maidenhair tree (*Ginkgo*). Angiospermae : with ovules enclosed in an ovary and seeds in a fruit, pollination being indirect through pollen-tubes growing down through stigmata and styles to the ovules in the ovaries. These are the " flowering plants " dominant in modern floras.

The following list of text-books is recommended for naturalists working on plants of the British flora :

General :

Fritsch, F. E., and Salisbury, E. J. (1943). Plant Form and Function. London, A. Bell & Sons.

Vascular Plants :

Bentham, G., and Hooker, J. D. (1924). Handbook of the British Flora (7th edition, ed. A. B. Rendle). London, L. Reeve & Co.

Fitch, W. H., and Smith, W. G. (1924). Illustrations of the British Flora (7th edition). London, L. Reeve & Co.

Hooker, J. D. (1930). The Student's Flora of the British Isles (3rd edition, reprinted). London, Macmillan.

Butcher, R. W., and Strudwick, F. E. (1944). Further Illustrations of British Plants (2nd edition). Ashford, L. Reeve & Co.

Druce, G. C. (1930). Hayward's Botanist's Pocket-book (19th edition). London, A. Bell & Sons.

Babington, C. C. (1922). Manual of British Botany (10th edition, ed. A. J. Wilmott). London, Gurney & Jackson.

Hyde, H. A., and Wade, A. E. (1940). Welsh Ferns. Cardiff, Nat. Museum of Wales.

Hyde, H. A. (1935). Welsh Timber Trees (2nd edition). Cardiff, Nat. Museum of Wales.

Bryophyta :

MacVicar, S. M. (1926). The Student's Handbook of British Hepatics (2nd edition). Eastbourne, V. V. Sumfield.

Dixon, H. N. (1924). The Student's Handbook of British Mosses (3rd edition). Eastbourne, V. V. Sumfield.

Lichenes :

Lorrain Smith, A. (1921). A Handbook of British Lichens. London, British Museum.

Fungi :

Ramsbottom, J. (1923). A Handbook of the Larger British Fungi. London, British Museum.

Massee, G. (1891). British Fungi. London.

Algae :

West, G. S., and Fritsch, F. E. (1927). A Treatise on the British Freshwater Algae. Cambridge, University Press.

Newton, L. (1931). A Handbook of British Seaweeds. London, British Museum.

Bacteria :

Partridge, W. (1946). Aids to Bacteriology (7th edition). London, Baillière, Tindall & Cox.

Students requiring more information on manuals, handbooks, etc. should consult *Bibliography of Key Works for the Identification of the British Fauna and Flora*, London, 1942 (published by the Systematics Association, 7*s*. 6*d*.).

APPENDIX C

BIOMETRICS

IT IS frequently essential to express characters in numerical terms for convenience of comparison, test of reliability, and other reasons. Even in simple descriptions, terms like "longer" and "shorter" or "larger" and "smaller" have no precise meaning except against a known standard, and very often are much better replaced by actual measurements. The application of measurements and statistical methods to plants and animals is termed biometrics. In its more advanced phases it requires a good knowledge of mathematics and the use of calculating machines. Naturalists wishing to obtain most useful results from their studies are advised to learn some of the elementary principles of biometrics and to use at least a minimum of the simpler formulae. The following rules, terms, and formulae will be found particularly valuable :

GENERAL RULES.

1. Plan the research carefully, well ahead of making time-consuming measurements (including counts and weights) and calculations.

2. Collect the material or data fairly by random sampling within the set limits. These last may be a population of a certain age, one growing on a certain soil-type or with certain exposure, or of a certain variety within a species, or a family bred from known parents, and so on.

3. The sample must be large enough, or given characters may be measured in every individual of the population (in theory, this is still a sample).

4. Measure and record the results ("score") to a stated degree of accuracy (say within 1 mm., 1 cm., or 1 gm.).

TECHNICAL TERMS, SYMBOLS, AND FORMULAE.

Variable. Any "character" measured, e.g. height of stem or number of flowers.

Arithmetic mean. This is simply :

$$\text{mean} = \frac{\text{sum of observations}}{\text{number of observations}}$$

Frequency (f) distribution. This is a grouping of all the measurements taken into classes (which may be given arbitrary limits).

Median. This is the middle observation. It is found by ranking observations in order of size and finding the central observation (or the mean of two in the middle if the number of observations is even).

Mode. This is the most frequently occurring observation, or in a frequency distribution table the class with the largest number of observations (items of measurement) in it.

Symbols. x or y = any single measurement.
$\bar{x}$ or $\bar{y}$ = mean.
N = total number of observations in a sample.
S or Σ = the sum of (as a rule, S is used for summation of sample values and Σ for derived quantities).

Standard deviation (σ or S.D.)

$$\text{Formula :} \quad \sigma = \sqrt{\frac{S(x-\bar{x})^2}{N}}$$

This formula gives the " scatter " of the observations around their mean (or average). Its importance lies in the fact that two samples may have the same mean but be very different in other respects for the same character. For example, in one sample a majority of observations may be very close to the mean while in another they may be widely scattered on both sides of it.

Standard error of mean

$$\text{Formula :} \quad \text{S. e. of mean} = \frac{\sigma}{\sqrt{N}}$$

This value is not in any sense a correction of mistakes made in calculations : it is simply an estimate of the probability that the result applies to the population from which the sample was drawn. It gives no information about the sample but about the (indefinitely large) population sampled.

Correlation (r)

$$\text{Formula :} \quad r = \frac{S(x-\bar{x})(y-\bar{y})}{N\sigma_x\sigma_y}$$

This is used when it is desired to know the degree of association between two variables. For example, in a given variety of plant is length of leaves correlated with breadth, are longer leaves also broader than short leaves ? Complete positive correlation gives the value 1 and complete negative correlation −1.

The following are very useful reference books for those wishing to understand more of the subject :

Chambers, E. G. (1940). Statistical Calculation for beginners. Cambridge, University Press.

Simpson, C. G., and Roe, A. (1939). Quantitative Zoology. New York and London, McGraw-Hill.

Tippet, L. H. C. (1937). The Methods of Statistics (2nd edition). London, Williams & Norgate.

Fisher, R. A. (1946). Statistical Methods for Research Workers (10th edition). Edinburgh and London, Oliver and Boyd.

APPENDIX D

ECOLOGICAL METHODS

ONE OF the many interesting features of ecology is that every ecologist for every new piece of research has more or less to invent his own methods. This statement is, perhaps, too wide but it is strictly true to say that the subject has so many aspects and, along modern lines, is so young that there is full opportunity for devising original methods and ingenious apparatus. This is particularly true for experimental work which opens up many possibilities for invention. Here only a few widely modifiable methods are outlined, but they have proved their general usefulness.

Recording. The general rule is to record all observations and experimental results (positive and negative) in writing at the time. For repeated scoring (observing, measuring, etc., and recording) it is often useful to prepare "score sheets" beforehand since they have the double advantage of preventing gaps and making final calculations easier. Squared graph paper will be found very useful.

Apparatus. A dozen white tapes marked at decimetre intervals, with 2 cm. extra at each end, of 1 m. marked length. 6 tapes marked at 1 m. intervals and 10 m. in length. Two dozen meat skewers. Pocket soil-testing outfit. Watkins Bee Meter. Pocket lens. Tins. Plant press. Vasculum. Ordnance Survey maps (1-inch and 6-inch).

Listing. Full and careful lists of all species in the area with notes on their frequency (symbols: d = dominant; a = abundant; f = frequent; o = occasional; r = rare; vr = very rare), stratification, life-forms, seasonal behaviour, and relationship to the varying features of the environment.

Quadrats. Areas, chosen either at random or as typical, are marked out with white tapes as 1 m. squares. A list quadrat is made by recording by name every species in the quadrat, with or without the number of individuals; a chart quadrat is made by recording to scale on squared paper every species, by letter or other symbols, in its correct position.

Transects. A tape is laid through a portion of a community and every plant touching the tape is recorded, usually on squared paper to scale by letter symbols. A belt transect is marked out by two parallel tapes and the plants between them listed or charted. A stratum transect records the heights of the plants touching the transect tape. A bisect is a stratum transect chart which includes underground systems as well as aerial shoots.

The investigator will find that two lines of research in the ecology of British plants are likely to yield particularly interesting results and to prove very stimulating to devising methods of study: the influence of biotic factors on plant life and the detailed study of single species or single plants (autecology). There is a very great deal to be done in both subjects. Those who are particularly attracted to the study of ecology might consider joining the British Ecological Society: Botanical Secretary, Prof. A. R. Clapham, Department of Botany, The University, Sheffield, 10. A simple guide to ecological methods is A. G. Tansley's *Practical Plant Ecology*, Allen & Unwin, 1923. Numerous papers on the ecology of British plant life (many containing details of new methods) have been published in *The Journal of Ecology*, Cambridge (1913-).

APPENDIX E

CYTOLOGY

(basic outline)

CYTOLOGY is the study of the structure of cells and the behaviour of their parts. A cell consists of a unit of protoplasm, a chemico-physical complex which is the basis of living activities, with various inclusions and, in plants, most often surrounded by a cell-wall. The protoplasm consists of two main parts : the cytoplasm or general and, assumedly, less specialized portion, and the nucleus, which in the non-dividing condition is clearly separated from the cytoplasm by a nuclear membrane. The detailed investigation of cells, their contents and activities is a laboratory process with a technique of its own and involving the use of special apparatus, in particular high-power compound microscopes. The subject has extended enormously in recent years, is still doing so, and has developed theories and terms of its own, some of which change or are added to with rather bewildering and perhaps unnecessary rapidity. Here it is intended to give only a bare outline of a few features of cell structure to elucidate terms and concepts used in the text (in particular in Chapter 12).

All living organisms at some time in a full life-history consist of a single cell. Unicellular plants and animals have the mature body of one cell only. In some groups (e.g. some Algae, Bryophyta, and Pteridophyta) there are two phases in the life-history in which the plant is unicellular, with intervening multicellular phases. In all except unicellular plants (and animals) there is multiplication of cells within one body or soma and the component cells of the body are termed somatic cells in distinction to the sexual cells or gametes. At ordinary somatic cell division one cell divides into two and division of the nucleus precedes division of the cytoplasm. At division of the nucleus there are present a definite number of distinct bodies known as chromosomes, constant as a rule for the species or variety of plant not only in number but also in size and shape. Every chromosome separates longitudinally into two equal halves (daughter chromosomes), which pass to opposite poles of a " spindle " and the two sets reform in two daughter nuclei. Nuclear division is followed by a separation of the cytoplasm into two parts bounded by a cell wall. The important point for theories of chromosome inheritance is the exact reduplication of the chromosome material in every body cell derived from the original one-celled fertilized egg (zygote) and (in some plants) spore.

Let us now consider the chromosome history in an ordinary flowering plant. The body (somatic) cells have nuclei with a fixed number of chromosomes and this number (with few exceptions) is even and is usually symbolized as "2n" (diploid). Thus: in *Ranunculus acris* 2n = 14, in *Silene maritima* 2n = 24, in *Centaurea scabiosa* 2n = 20, and in *C. jacea* 2n = 44. Throughout the numerous cell divisions occurring in the roots, stems, leaves, and flowers, excluding only those involved in the formation of the gametes and a few associated cells, the somatic number is repeated again and again. The pollen mother-cell and the embryo-sac mother-cell undergo two divisions in rapid succession so that the resulting pollen-grain nuclei and the 4 (potential) embryo-sac nuclei have only half the number of chromosomes of the body cells, a number symbolized by "n" (haploid). The pollen-cell nucleus gives rise by ordinary nuclear division to a pollen-tube nucleus and two male gametes or sperms. Typically all four cells (tetrad) derived from a pollen mother-cell produce functioning pollen-grains, while only one cell derived from an embryo-sac mother-cell gives rise to a functioning embryo-sac, in which a number of ordinary nuclear divisions produce eight free cells of which one is a female gamete or egg and two of the others fuse together to form what is termed a "secondary nucleus." There are many modifications of this scheme, but the important points are (1) a reduction of chromosome numbers prior to the formation of the gametes (male and female sex cells) from 2n to n, from diploid to haploid, and (2) the formation, in the embryo-sac, of a diploid (2n) "secondary nucleus."

In the flowering plant, fertilization follows the reduction division relatively quickly. Pollination, production of a pollen-tube, and movement down the latter of the male gametes bring one male gamete into contact with the female gamete of an ovule, and the two fuse. The second male gamete also enters the embryo-sac and fuses with the secondary nucleus. There thus result a fertilized egg (or zygote) and a fertilized secondary nucleus (or endosperm nucleus). The zygote has 2n chromosomes (it is diploid); the endosperm nucleus has 3n chromosomes (it is triploid). There is a 2 : 3 ratio in chromosome number between the zygote (developing by ordinary cell division into the embryo) and the endosperm nucleus (giving rise by ordinary cell division to the endosperm). If, for any reason, this ratio is upset, a viable seed may not form and this can be one cause of sterility.

There is a risk of some confusion in the use of the words "reduction," "halving," "doubling" and so on in reference to the reproductive process. It is the diploid number which is reduced or halved but the haploid number which is doubled at fertilization. It is, therefore, more precise to use the terms "mitosis," for ordinary nuclear divisions in which the chromosome number remains unchanged, and "meiosis," for the two divisions resulting in reduction or halving of the chromosome number relative to the number in the preceding somatic nuclei. Further complications, however, occur. Occasionally gametes are formed with more than one chromosome set and if they take part in fertilization, plants with more than two chromosome sets may result. These latter are termed polyploids, and polyploidy has been of considerable importance in the evolution of the different kinds of existing plants, though its long-range evolutionary importance is more difficult to estimate. It is, therefore, advisable to refer to a "basic" chromosome number for a group and to symbolize this as "x," while the symbols : n, 2n, 3n, etc., refer to the actual chromosome constitution in a given life-history. Unfortunately, the terms haploid (one), diploid (two), triploid (three), tetraploid (four) are used relatively with reference sometimes to x and sometimes (more often) to n. Since the value of n is definitely ascertainable by observation, while that of x has to be deduced, it is best to use the -ploid terms with reference to n, at least unless another relativity is clearly stated. Polyploidy can also arise by doubling of chromosomes in somatic cells.

Our example of meiosis and fertilization was a generalized flowering plant (Angiosperm). In all sexually reproducing plants the full life-history includes these two processes but there are interesting differences in the immediate results. Two examples must suffice. In a moss, meiosis occurs at the formation of the spores in the young capsule. These spores on germination give rise to threads or, more rarely, thin plates of cells from which the shoots of stem and leaves arise. On these shoots the reproductive organs with gametes are produced. Every cell from the spore to gametes inclusive has the haploid (n) chromosome number in its nucleus. Fertilization gives a diploid (2n) zygote (fertilized egg) which grows out into stalk and capsule with diploid (2n) nuclei. Here there is an alternation of generations with the gametophyte generation the important vegetative as well as the sexual phase. All the characters of the vegetative parts and of the male and female organs are correlated with the haploid condition of the nuclei. In a fern, there is also a marked alternation of generations, but here the spore gives rise to a small plate or cushion of cells, with no differentiation into stem and leaves, and termed the prothallium, on which the male and female organs arise. The fertilized egg grows at once into a sporeling and this develops into the mature plant with stem and leaves (or fronds). On the latter, or a modified form of them, arise the spore-cases inside which spores are formed, meiosis occurring as the spore mother-cells form tetrads of spores.

Thus we have the schemes :

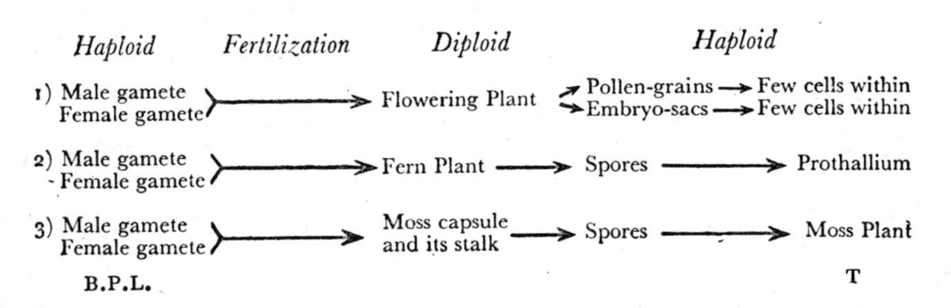

Reference works:

Sharp, L. W. (1934). An Introduction to Cytology (3rd edition). New York and London, McGraw-Hill.

Darlington, C. D. (1932). Chromosomes and Plant-Breeding. London, Macmillan.

Darlington, C. D. (1937). Recent Advances in Cytology (2nd edition). London, J. & A. Churchill.

Darlington, C. D., and La Cour, L. F. (1942). The Handling of Chromosomes. London, Allen & Unwin.

Maude, P. F. (1939). The Merton Catalogue, a list of the chromosome numbers of species of British Flowering Plants. Cambridge, University Press (*New Phytologist*, Reprint No. 20).

APPENDIX F

GENETICS
(basic outline)

GENETICS is the study of heredity. In present practice it is mainly an extension of the methods employed by Mendel (1822–1884), with garden peas, and the correlation of these with cytology. According to the views now widely accepted, hereditary factors or genes are located in the chromosomes more or less like beads on a string. These chromosomal genes are under the control of the normal mechanisms of mitosis and meiosis, though in varying ways and degrees they may break away occasionally from such control. Genes connected with the plastids or with the cytoplasm do not follow the same rules as those of the chromosomes. We concern ourselves in this outline with the last mentioned.

Genes are often referred to as determiners of characters (of structure or of function). It is often convenient to speak of a gene as determining a character, and earlier in the study of Mendelian genetics it was thought that a unit gene was the internal determiner of a unit character. It may now be doubted if that is ever a true picture. Apart from little-understood interactions with their cytoplasmic and more external environments, genes interact in complicated ways with one another. Only a few of the simpler examples of such interaction can be given here in a later paragraph. It does, however, remain a fact that some genes have so pronounced an influence, under certain usual conditions at least, in determining a definable character that the study of their behaviour in crosses enables us to formulate rules of heredity, which have a wide application, and by the laws of chance to give a mathematical expression to the results of inbreeding and outbreeding in so far as such genes and characters are concerned.

The simplest results are obtained when paired contrasting characters are involved. Suppose, for example, we choose the sea bladder campion (*Silene maritima*) and consider first the seed-coat markings designated tubercled and armadillo (see p. 201). To save space we designate "tubercled" by **T** and "armadillo" by **t**. All plants with armadillo seeds "breed true" to this character, that is all offspring from self-pollinations, or from cross-pollinations between two plants with armadillo seeds, have armadillo seeds. On the other hand, some plants with tubercled seeds breed true and some do not. If either of the two crosses be made :

T(as ovule parent) × **t** (as pollen parent) or
t (as ovule parent) × **T** (as pollen parent)

all the offspring have tubercled (T) seeds. In other words T is dominant and t is recessive in first or F_1 generations from the reciprocal crosses. If F_1 plants are selfed or one crossed with another and a second (F_2) generation obtained it is found that both tubercled and armadillo plants occur amongst the offspring and in the ratio of 3T : 1t, or approximating to this, if sufficiently large numbers are scored. Remembering that fertilization is the union of a female and a male gamete (it is conventional to put the female first—"ladies first") and that a zygote and the plant derived from it are of biparental origin (in this sense at least) we can symbolize thus :

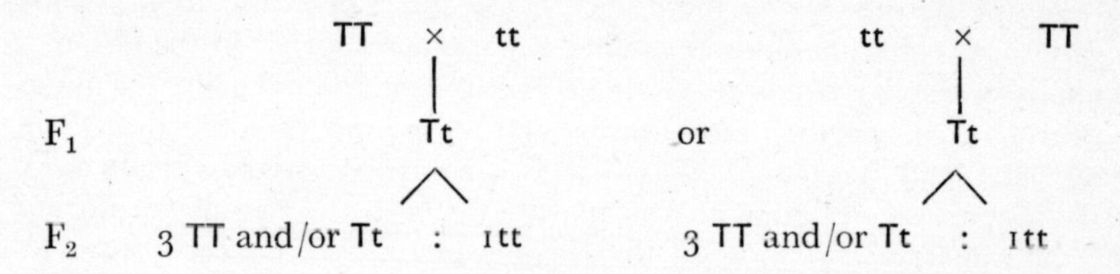

It must be emphasized that tubercled seeds of TT plants are indistinguishable from those of Tt plants. The former breed true and are said to be homozygous for this character, while the latter segregate on selfing or crossing together and are said to be heterozygous. It is important to realize that plants may be alike in appearance (or phenotypically similar) yet unlike in the offspring they give (genotypically dissimilar). Plants with the recessive gene in double dose (tt) are always homozygous and phenotypically and genotypically similar for this character. We also note that the armadillo character disappeared from the F_1 generation but reappeared (segregated out) in F_2. An exactly comparable result is obtained by crossing a plant with full-sized petals with one having poor petals, the former character being dominant and the latter recessive.

Dominance, however, is not always to be observed, especially in dealing with characters of wild plants. An interesting example in *Silene* is hairiness (indumentum). *Silene maritima* is always glabrous but the common bladder campion (*S. cucubalus*) has a hairy variety. If a densely hairy *S. cucubalus* be crossed with *S. maritima* the F_1 plants are somewhere in between the two extremes in hairiness. These selfed or bred together give a ratio of 1 densely hairy : 2 intermediate : 1 glabrous. Using the symbols H for dense indumentum and h for glabrous, we can tabulate the results as :

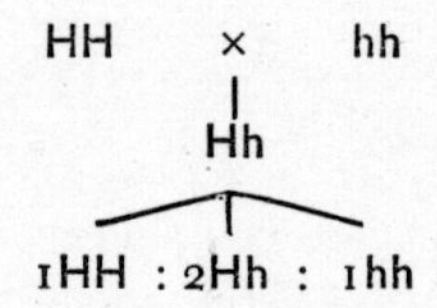

The above results, whether dominance occurs or not, are understandable if we recall the behaviour of chromosomes at mitosis and meiosis and the fundamentals of fertilization. In ordinary (somatic) nuclear division every chromosome becomes represented by an equivalent in the offspring nuclei so that throughout the body the nuclei have all qualitatively and quantitatively equal sets of chromosomal genes. Since the first body-cell (zygote) originated at fertilization by fusion of a female with a male gamete, it has two sets of chromosomes. At meiosis this number is halved and this reduction also involves a segregation of genes carried in the same position (locus) on two equivalent chromosomes. If the corresponding genes at a locus are different the gametes will to that extent be different in their chromosomal genes. For instance, in the F_1 of *Silene maritima* from the cross T×t, we designated the body constitution as Tt. Accepting these letters as symbols for genes, in meiosis of the pollen mother-cell and formation of a tetrad of four viable pollen-grains, two of these would have T and two t. A corresponding segregation of genes occurs on the ovule side (though only one of the potential embryo-sac mother-cells functions). Thus the body can be heterozygous but the gametes are " pure."

We can now plan out the results of breeding experiments of which we have outlined a few examples in a chequer-board arrangement, which is often a most useful, though long-hand, way of understanding or foretelling results. In order to generalize further we will use A and a and B and b as symbols for any two pairs of characters, or the genes which are apparently principal determiners for the characters. Such pairs are termed allelomorphs or alleles. Suppose crosses to be made as follows: (1) AA×aa ; (2) BB×bb; (3) AABB×aabb ; we lay out the results in chequer-board fashion thus for F_2 generations :

I

	A	a
A	AA	Aa
a	aA	aa

Ratios : 1AA : 2Aa : 1aa
from selfing Aa or crossing two Aa plants together.

II

	B	b
B	BB	Bb
b	bB	bb

1BB : 2Bb : 1bb
from selfing Bb or crossing two Bb plants together.

If A be dominant over a, and B over b, the empirical ratios will be 3A : 1a and 3B : 1b.

	AB	Ab	aB	ab
AB	AB AB 1	AB Ab 2	AB aB 3	AB ab 4
Ab	Ab AB 5	Ab Ab 6	Ab aB 7	Ab ab 8
aB	aB AB 9	aB Ab 10	aB aB 11	aB ab 12
ab	ab AB 13	ab Ab 14	ab aB 15	ab ab 16

Ratio : 1 AABB : 1 AAbb : 1 aaBB : 1 aabb : 2 AABb : 2 AaBB : 2 Aabb : 2 aabB : 4 AaBb ;
from selfing AaBb or crossing two AaBb plants together.

If A be dominant over a and B over b then the empirical ratio will be 9AB : 3Ab : 3aB : 1ab. The important points to note are (1) that plants 1, 6, 11, 16 are homozygous for both characters ; (2) that plants 4, 7, 10, 13 are heterozygous for both characters ; (3) that the remaining eight plants are heterozygous for one pair and homozygous for the other pair ; (4) plants 6 and 11 are new true-breeding combinations, different from the two grandparents.

If the student is at any time puzzled by formulae or by actual breeding results it will be found very helpful to plot out the possibilities on a chequer-board scheme. First determine all the possible combinations of the gametes, assuming the rule of " the purity of the gametes," that only one of any one pair of allelomorphic genes (A *or* a, B *or* b, not both A *and* a or B *and* b) can occur together in a gamete. Make out a " chequer-board " with a number of squares equal to the square of the ascertained number of kinds of gametes. Write out the gametes along the top and down the left-hand side of the chequer-board to represent female and male gametes respectively, one opposite each tier or row of squares. Then fill in the combination in each square by inserting the female and male gamete constitution (for the usual conveniences of the scheme the order of the genes in any one square, representing a zygote, does not matter). In preparing the normal chequer-board scheme, resolving the symbols, and summating the results, the following assumptions are at first made :

1. that any other gene combination than allelomorphs can occur in the gametes, thus A *and* B, a *and* B, and so on may occur together ;

2. that the different possible kinds of gametes are all viable and occur in equal numbers ;

3. that fertilization (and antecedent pollination) is entirely at random so far as the gene composition of the gametes is concerned ;

4. that the postulated genes are independent of one another in their determination of the characters scored ;

5. that all zygotes survive and give viable seeds and scorable seed-lings.

If ratios widely different from those expected occur they must be examined to see if they can be explained through one or more of the above assumptions not being true. For example, in self-incompatibility (see p. 204) fertilization is not at random since the rate of growth of pollen-tubes through stylar tissue depends on relations between the gene constitution of the pollen and style nuclei. Or again, it may be necessary for two (or more) genes to occur together in somatic nuclei for a given character to appear. Thus a certain colour is often produced only when two genes are present together, either alone giving absence of that colour. There are many known examples of departure from simple " Mendelian ratios." Some of these have been interpreted in line with the main underlying theory of inheritance outlined above, with modifications in details. Whether these modifications have now made the main theory itself top-heavy and the time has come to seek some more general, all-inclusive theory cannot be discussed here. In particular, attention may be called to an increasing recognition of the importance of the interaction of genes, of the occurrence of large numbers of genes with small but cumulative effects, and of genes (if so they be called) outside the nucleus in the plastids or the cytoplasm.

One other matter needs explanation. In general, within a species or variety there is a definite and constant chromosome number. This number is less, frequently very much less, than the number of genes that has to be postulated to explain breeding results. It follows that a single chromosome carries more than one, and usually very many, genes. In other words, in any one chromosome there are a number, often a large number, of gene loci. Genes with loci in one and the same chromosome are said to be linked and this linkage expresses itself in the results of breeding through characters remaining associated, or conversely remaining unassociated, when they might be expected to show random association and dissociation. This linkage can, however, be broken when, at the reduction division, pairs of chromosomes (homologues) twine round each other and exchange parts. This process of "crossing-over," as it is termed, is of great importance in modern theoretical genetics, because, on the reasonable assumption that an exchange is more likely to occur between two gene loci farther apart than between two closer together, the relative positions of gene loci have been calculated and maps of the chromosomes prepared.

Books recommended include :

Ford, E. B. (1938). The Study of Heredity. London, Thornton & Butterworth.
Ford, E. B. (1934). Mendelism and Evolution (2nd edition). London, Methuen.
Huxley, J. (1942). Evolution, the Modern Synthesis. London, Allen & Unwin.
Ford, E. B. (1945). Butterflies. London, Collins (NEW NATURALIST).

APPENDIX G

THE MORE IMPORTANT COUNTY AND LOCAL FLORAS

Berkshire : Druce, G. C. (1897). The Flora of Berkshire. Oxford, Clarendon Press.

Bristol : White, J. W. (1912). The Flora of Bristol. Bristol, John Wright and Sons.

Buckinghamshire : Druce, G. C. (1926). The Flora of Buckinghamshire. Arbroath, T. Buncle & Co.

Cambridgeshire : Evans, A. H. (1939). A Flora of Cambridgeshire. London, Gurney & Jackson.

Cheshire : Tabley, Lord de (1899). The Flora of Cheshire. London, Longmans, Green & Co.

Cornwall : Davey, F. H. (1909). Flora of Cornwall. Penryn, F. Chegwidden.

Cumberland : Hodgson, W. (1898). Flora of Cumberland. Carlisle, W. Meals & Co.

Derbyshire : Linton, W. R. (1903). Flora of Derbyshire. London, Bemrose & Sons, Ltd.

Devon : Martin, W. K., and Fraser, G. T. (1939). Flora of Devon. Arbroath, T. Buncle & Co.

Dorset : Mansel-Pleydell, J. C. (1895). The Flora of Dorsetshire (2nd edition). Dorchester, " Dorset County Chronicle " Printing Works.

Hampshire : Townsend, F. (1904). Flora of Hampshire (2nd edition). London, Lovell Reeve & Co., Limited.

Bournemouth : Linton, E. F. (1919). Flora of Bournemouth (2nd issue). Bournemouth (no publisher given).

Herefordshire : Purchas, W. H., and Ley, A. (1889). A Flora of Herefordshire. Hereford, Jakeman & Carver.

Hertfordshire : Pryor, A. R. (1887). A Flora of Hertfordshire. London, Gurney & Jackson.

Kent : Hanbury, F. J., and Marshall, E. S. (1899). Flora of Kent. London, Frederick J. Hanbury.

Lancashire, West : Wheldon, J. A., and Wilson, A. (1907). The Flora of West Lancashire. Eastbourne, W. T. Sumfield.

Leicestershire and Rutland : Horwood, A. R., and Noel, C. W. F. (1933). The Flora of Leicestershire and Rutland. Oxford, University Press.

Liverpool District : GREEN, C. T. (1933). The Flora of the Liverpool District (5th edition). Arbroath, T. Buncle & Co.

Middlesex : TRIMEN, H., and DYER, W. T. T. (1869). Flora of Middlesex. London, Robert Hardwicke.

Norfolk : NICHOLSON, W. A. (1914). A Flora of Norfolk. London, West, Newman & Co.

Northamptonshire : DRUCE, G. C. (1930). The Flora of Northamptonshire. Abroath, T. Buncle & Co.

Oxfordshire : DRUCE, G. C. (1927). The Flora of Oxfordshire (2nd edition). Oxford, Clarendon Press.

Somerset : MURRAY, R. P. (1896). The Flora of Somerset. Taunton, Barnicott & Pearce.

Suffolk : HIND, W. M. (1889). The Flora of Suffolk. London, Gurney & Jackson.

Surrey : SALMON, C. E. (1931). Flora of Surrey. London, G. Bell & Sons, Ltd.

Sussex : WOLLEY-DOD, A. H. (1937). Flora of Sussex. Hastings, Kenneth Saville.

Warwickshire : BAGNALL, J. E. (1891). The Flora of Warwickshire. Birmingham, Cornish Brothers.

Westmorland : WILSON, A. (1938). The Flora of Westmorland. Arbroath, T. Buncle & Co.

Wiltshire : PRESTON, T. A. (1888). The Flowering Plants of Wilts. Wiltshire Archaeological and Natural History Society.

Worcestershire : AMPHLETT, J., and REA, C. (1909). The Botany of Worcestershire. Birmingham, Cornish Brothers, Ltd.

Yorkshire, East Riding : ROBINSON, J. F. (1902). The Flora of the East Riding of Yorkshire. London, A. Brown & Sons, Ltd.

Yorkshire, North : BAKER, J. G. (1906). North Yorkshire (2nd edition). London, A. Brown & Sons, Ltd.

Yorkshire, West : LEES, F. A. (1888). The Flora of West Yorkshire. London, Lovell Reeve & Co.

Anglesey and Carnarvonshire : GRIFFITH, J. E. (1895). The Flora of Anglesey and Carnarvonshire. Bangor, Nixon & Jarvis.

Cardiganshire : SALTER, J. H. (1935). The Flowering Plants and Ferns of Cardiganshire. Cardiff, University Press Board.

Glamorgan : TROW, A. H. (1911). The Flora of Glamorgan. Cardiff, Cardiff Naturalists' Society.

Dumfriesshire : SCOTT-ELLIOT, G. F. (1896). Flora of Dumfriesshire. Dumfries, J. Maxwell & Son.

Perthshire : WHITE, F. B. W. (1898). The Flora of Perthshire. Perthshire Society of Natural Science.

Orkney : SPENCE, M. (1914). Flora Orcadensis. Kirkwall, D. Spence.

Uig (Lewis) : Campbell, M. S. (1945). The Flora of Uig. Arbroath, T. Buncle & Co.

Dublin : Colgan, N. (1904). The Flora of the County Dublin. Dublin, Hodges, Figgis & Co. Ltd.

Donegal : Hart, H. C. (1898). Flora of the County Donegal. Dublin, Sealy, Bryers & Walker.

Ireland, North-East : Stewart, S. A., and Corry, T. H. (1938). A Flora of the North-East of Ireland (2nd edition). Belfast, The Quota Press.

Kerry : Scully, R. W. (1916). Flora of County Kerry. Dublin, Hodges, Figgis & Co., Ltd.

Guernsey: Marquand, E. D. (1901). Flora of Guernsey. London, Dulau & Co.

Jersey: Lester Garland, L. V. (1903). A Flora of the Island of Jersey. London, West, Newman & Co.

REFERENCES

THE LIST of references given below is intended to serve two purposes: (1) to suggest further reading to students wishing to know more about subjects only outlined in this book, and (2) to provide a check to information given in the text, especially when this has been based on recently published material. It does not pretend to be a complete guide to the literature on British plant life. The references for Chapters 8 and 9 have been very considerably reduced, because a long bibliography on the ecology of British plants is given in (141). The references for Chapter 12 are relatively very numerous, because there is no published book (or bibliography) dealing in a general manner with the genetics of British plants. A considerable number of references in (153) are relevant to Chapters 1 and 14. Many of the books and papers have longer or shorter lists of references and attention is called also to the works quoted in the Appendixes.

Chapter 1

(34), (136), (153), (177). [General introductory reading]

Chapter 2

(5), (27), (62), (68), (129), (132), (181). [Books on early history of the earth and its inhabitants]

Chapter 3

(10), (24), (27), (114), (126), (127), (128), (129), (153), (181). [Publications on the early history of land plants]

Chapter 4

(27), (112), (113), (129), (134), (135), (159), (172), (179), (180), (181). [Works on the influence of the Ice Age]

Chapter 5

(2), (20), (29), (30), (34), (36), (37), (38), (39), (112), (113), (122), (134), (135), (159), (166), (172), (179), (180), (181). [References to post-glacial history of the British Flora]

Chapter 6

(12), (40), (50), (51), (52), (53), (88), (88a), (108), (109), (121), (134),

(135), (147), (148), (163), (164), (173). [Accounts of the present composition of the British Flora]

Chapter 7

(40), (88), (88a), (108), (109). [References concerning the geographical relationships of the British Flora]

Chapter 8

(11), (14), (18), (78), (118), (119), (120), (138), (140), (141). [Selected references to ecological works]

Chapter 9

(1), (13), (55), (56), (104), (105), (117), (118), (120), (140), (141), (146), (157). [Selected references to ecological works]

Chapter 10

(35), (49), (59), (60), (77), (78), (149), (150), (154), (156), (178). [Works relevant to the study of variation in British plants]

Chapter 11

(3), (13), (35), (47), (59), (60), (61), (72), (77), (110), (111), (115), (117), (118), (131), (136), (149), (155), (156), (157), (182). [Accounts connected with adaptation and natural selection]

Chapter 12

(4) *Trifolium*, (6) *Melandrium*, (7) *Rosa*, (8) *Salix*, (9) *Alchemilla*, (15) *Viola*, (16) *Viola*, (17) *Viola*, (21) *Lamium*, (22) *Rubus*, (23) *Rubus*, (25) *Primula*, (26) *Taraxacum*, (28) *Lotus*, (31) *Trifolium*, (32) *Lythrum*, (33) *Viola*, (41) *Lolium*, (42) *Rubus*, (43) *Rubus*, (44) *Rosa*, (45) *Rosa*, (46) *Rosa*, (48) *Salix*, (54) *Carex*, (57) *Nasturtium*, (58) *Rosa*, (63) *Lilium*, (64) *Lolium*, (65) *Lolium*, (66) *Lolium*, (69) *Bryonia*, (70) *Lamium*, (71) *Digitalis*, (73) *Epilobium*, (74) *Melandrium*, (75) *Geum*, (76) *Ranunculus*, (77) *Silene*, (79) *Ranunculus*, (80) *Saxifraga*, (81) *Centaurea*, (82) *Primula*, (83) *Anthyllis*, (84) *Saxifraga*, (85) *Centaurea*, (86) *Saxifraga*, (87) *Primula*, (89) *Ranunculus*, (90) *Epilobium*, (91) *Lamium*, (92) *Lamium*, (93) *Galeopsis*, (94) *Lamium*, (95) *Galeopsis*, (96) *Galeopsis*, (97) *Potentilla*, (98) *Galeopsis*, (99) *Potentilla*, (100) *Galeopsis*, (101) *Papaver*, (102) *Silene*, (103) *Salix*, (106) *Saxifraga*, (107) *Taraxacum*, (116) *Centaurea*, (123) *Silene*, (124) *Digitalis*, (125) *Digitalis*, (130) *Valeriana*, (133) *Melandrium*, (137) *Lythrum*, (139) *Rosa*, (143) *Senecio*, (144) *Senecio*, (145) *Alchemilla*, (151) *Taraxacum*, (158) *Viola*, (160) *Carex*, (162) *Melandrium*, (165) *Ranunculus*, (167) *Salix*, (168)

Trifolium, (169) *Trifolium*, (170) *Trifolium*, (171) *Trifolium*, (174) *Melandrium*. (175) *Papaver*, (176) *Erophila*. [Publications concerning heredity in the genera named]

Chapter 13

(19), (29), (67), (142), (161). [References concerning the present and future of British plant life]

Chapter 14

(19), (59), (142), (153). [Books or papers for final reference]

BIBLIOGRAPHY

(1) ANDERSON, V. L. (1927). The water economy of the chalk flora. *J. Ecology, London, 15* : 72–129.

(2) Anon. (1940). Cultivated crops in early England. *Nature, London, 146* : 744.

(3) ARBER, A. (1920). Water Plants. Cambridge, University Press.

(4) ATWOOD, S. S., and KREITLOW, K. W. (1946). Studies of a genetic disease of *Trifolium repens* simulating a virosis. *Amer. J. Bot. 33* : 91–100.

(5) BACON, J. S. D. (1944). The Chemistry of Life. London, C. A. Watts.

(6) BAKER, H. G. (1943). Petal-colour inheritance in *Lychnis. Nature, 152* : 161–62.

(7) BLACKBURN, K. B., and HESLOP-HARRISON, J. W. (1921). The status of the British rose forms as determined by their cytological behaviour. *Ann. Bot. 35* : 159–88.

(8) (1924). A preliminary account of the chromosomes and chromosome behaviour in the Salicaceae. *Ann. Bot. 38* : 361–78.

(9) BÖÖS, G. (1924). Neue embryologische Studien über *Alchemilla arvensis* (L.) Scop. *Bot. Not.* 1924 : 209–50.

(10) BOWER, F. O. (1935). Primitive Land Plants. London, Macmillan.

(11) BRADE-BIRKS, S. G. (1944). Good Soil. London, English Universities Press.

(12) CAMPBELL, M. S. (ed.) (1945). The Flora of Uig (Lewis). Arbroath, Buncle.

(13) CHAPMAN, V. J. (1942). The new perspective in the halophytes. *Quart. Rev. Biol. 17* : 291–311.

(14) CLARKE, G. R. (1936). The Study of the Soil in the Field. Oxford, Clarendon Press.

(15) CLAUSEN, J. (1927). Chromosome number and the relationship of species in the genus *Viola. Ann. Bot. 41* : 677–714.

(16) (1931). *Viola canina* L., a cytologically irregular species. *Hereditas, Lund, 15* : 67–88.

(17) (1931). Cyto-genic and taxonomic investigations on *Melanium* violets. *Hereditas, Lund, 15* : 219–308.

(18) COMBER, N. M. (1936). An Introduction to the Scientific Study of the Soil. (3rd edition). London, Arnold.

(19) Committee Report. (1944). Nature conservation and nature reserves. *J. Ecology, London, 32* : 45–82.

(20) CONOLLY, A. P. (1941). A report of plant remains from Minnis Bay, Kent. *New Phyt. 40* : 299–303.

(21) CORRENS, C. (1926). Genetische Untersuchungen an *Lamium amplexicaule* L. *Biol. Zbl. Leipzig, 46 :* 67–79.

(22) CRANE, M. B., and LAWRENCE, W. J. C. (1931). Inheritance of sex colour and hairiness in the raspberry, *Rubus idaeus* L. *J. Genet. Cambridge, 25 :* 243–55.

(23) CRANE, M. B., and THOMAS, R. T. (1940). Reproductive versatility in *Rubus*. *J. Genet. Cambridge, 40 :* 109–28.

(24) CROOKALL, R. (1929). Coal Measure Plants. London, Arnold.

(25) CROSBY, J. L. (1940). High proportions of homostyle plants in populations of *Primula vulgaris*. *Nature, London, 145 :* 672.

(26) CURTIS, W. M. (1940). The structure and development of some apomicts of *Taraxacum*. *Kew Bull.* 1940 : 1–29.

(27) DARRAH, W. C. (1939). Principles of Palaeobotany. Leiden, *Chronica Botanica*.

(28) DAWSON, C. D. R. (1941). Tetrasomic inheritance in *Lotus corniculatus* L. *J. Genet. Cambridge, 42 :* 49–72.

(29) DRUCE, G. C. (1920). The extinct and dubious plants of Britain. *Bot Exch. Club 1919 Report*, 731–99.

(30) ERDTMAN, G. (1943). An Introduction to Pollen Analysis. Waltham, Mass., U.S.A., *Chronica Botanica*.

(31) ERITH, A. G. (1928). Some hybrids of varieties of white clover (*Trifolium repens*). *J. Genet. Cambridge, 19 :* 351–55.

(32) FISHER, R. A., and MATHER, K. (1943). The inheritance of style length in *Lythrum salicaria*. *Ann. Eugenics, 12 :* 1–23.

(33) FOTHERGILL, P. G. Studies in *Viola* :—I. *Genetica, 's Gravenhage, 20 :* 159–85 ; II. *ib. 21 :* 153–76 ; III. *New Phyt. 40 :* 139–51 ; IV. *ib. 43 :* 23–35.

(34) FOX, C. (1943). The Personality of Britain. Cardiff, National Museum of Wales.

(35) GATES, R. R. (1921). Mutations and Evolution. *New Phyt.* Reprint No. 12.

(36) GODWIN, H. (1943). Coastal peat beds of the British Isles and North Sea. *J. Ecology, London, 31 :* 199–247.

(37) (1944). Neolithic forest clearance. *Nature, London, 153 :* 511–12.

(38) GODWIN, H., and CONWAY, V. M. (1939). The ecology of a raised bog near Tregaron, Cardiganshire. *J. Ecology, London, 27 :* 313–59.

(39) GODWIN, H., and TANSLEY, A. G. (1941). Prehistoric charcoals as evidence of former vegetation, soil, and climate. *J. Ecology, London, 29 :* 117–26.

(40) GOOD, R. D'O. (1928). Notes on a comparison of the Angiosperm floras of Kent and Pas de Calais. *J. Bot. 66 :* 253–64.

(41) GREGOR, J. W. (1928). Pollination and seed production in the rye-grasses (*Lolium perenne* and *L. italicum*). *Trans. Roy. Soc. Edinb. 55* : 773–94.

(42) GUSTAFSSON, A. (1942). The origin and properties of the European blackberry flora. *Hereditas, Lund, 28* : 249–77.

(43) (1943). The genesis of the European blackberry flora. *Lunds Univ. Aarsskr. N.F. 39* : No. 6.

(44) (1944). The constitution of the *Rosa canina* complex. *Hereditas, Lund, 30* : 405–28.

(45) GUSTAFSSON, A., and HAKANSSON, A. (1942). Meiosis in some *Rosa* hybrids. *Bot. Not.* ; 1942. 331–43.

(46) GUSTAFSSON, A., and SCHRÖDERHEIM, J. (1945). Ascorbic acid in *Rosa* hybrids, *Hereditas, Lund, 31* : 489–97.

(47) HABERLANDT, G. (1914). Physiological Plant Anatomy (transl. M. Drummond). London, Macmillan.

(48) HAKANSSON, A. (1938). Zytologische Studien an *Salix*-Bastarden. *Hereditas, Lund, 24* : 1–34.

(49) HARRISON, J. W. HESLOP (1926). Heterochromosomes and polyploidy. *Nature, London, 117* : 50.

(50) (1939). Fauna and flora of the Inner and Outer Hebrides. *Nature, London, 143* : 1004–07.

(51) (1941). A preliminary flora of the Outer Hebrides. *Proc. Univ. Durham Phil. Soc. 10* : 228–73.

(52) (1945). Noteworthy sedges from the Inner and Outer Hebrides. *Trans. Proc. Bot. Soc. Edinb. 34* : 270–77.

(53) HARRISON, J. W. HESLOP, and BLACKBURN, K. B. (1946). The occurrence of a nut of *Trapa natans* L. in the Outer Hebrides. *New Phyt. 45* : 124–31.

(54) HEILBORN, O. (1924). Chromosome numbers and dimensions, species-formation and phylogeny in the genus *Carex*. *Hereditas, Lund. 5* : 129–216.

(55) HEPBURN, I. (1943). A study of the vegetation of sea-cliffs in North Cornwall. *J. Ecology, London, 31* : 30–39.

(56) HOPE-SIMPSON, J. F. (1940). Studies of the vegetation of the English Chalk :—VI. *J. Ecology, London, 28* : 386–402.

(57) HOWARD, H. W., and MANTON, I. (1946). Autopolyploid and allopolyploid watercress with the description of a new species. *Ann. Bot. N.S. 10* : 1–13.

(58) HURST, C. C. (1928). Differential polyploidy in the genus *Rosa* L. *Z. indukt. Abstamm.- u. VererbLehre*, 1928 Suppl. : 866–906.

(59) HUXLEY, J. S. (ed.) (1940). The New Systematics. Oxford, Clarendon Press.

(60) (1942). Evolution, the Modern Synthesis. London, Allen & Unwin.

(61) JAMES, W. O., and CLAPHAM, A. R. (1935). The Biology of Flowers. Oxford, Clarendon Press.

(62) JEANS, J. (1943). Evolution in astronomy. *Nature, London, 151* : 7.

(63) JENKIN, T. J. (1928–30). Inheritance in *Lolium perenne*. *J. Genet. Cambridge, 19* : 391–402 ; *ib. 19* : 404–17 ; *ib. 22* : 389–94.

(64) (1934). Self and cross-fertilization in *Lolium perenne* L. *J. Genet. Cambridge, 28* : 11–17.

(65) (1934-35). Interspecific and intergeneric hybrids in herbage grasses. *J. Genet. Cambridge, 28* : 205–64 ; *ib. 31* : 379–411.

(66) JENKIN, T. J., and THOMAS, P. T. (1938). The breeding affinities and cytology of *Lolium* species. *J. Bot. 76* : 10–12.

(67) Joint Communication (1935). Changes in the British fauna and flora during the past fifty years. *Proc. Linn. Soc., Session 148* : 33–52.

(68) JONES, H. S. (1940). Life on Other Worlds. London, English Universities Press.

(69) JONES, W. NEILSON, and RAYNER, M. C. (1916). Mendelian inheritance in varietal crosses of *Bryonia dioica*. *J. Genet. Cambridge, 5* : 203–24.

(70) JÖRGENSEN, C. A. (1927). Cytological and experimental studies in the genus *Lamium*. *Hereditas, Lund, 9* : 126–36.

(71) KEEBLE, F. C. PELLEW, and JONES, W. N. (1910). The inheritance of peloria and flower-colour in foxgloves (*Digitalis purpurea*). *New Phyt. 9* : 68–77.

(72) KNUTH, P. (190–609). Handbook of Flower Pollination (transl. J. R. A. Davis). Oxford, Clarendon Press. 3 vols.

(73) LEHMANN, E. (1939). Zur Genetik der Entwicklung in der Gattung *Epilobium*. *Jb. Wiss. Bot. 87* : 625–41 ; *ib. 88* : 284–343.

(74) LÖVE, D. (1944). Cytogenetic studies in dioecious *Melandrium*. *Bot. Not.* 1944 : 125–213.

(75) MARSDEN-JONES, E. M. (1930). The genetics of *Geum intermedium* Willd. *haud* Ehrh. and its back-crosses. *J. Genet. Cambridge, 23* : 377–95.

(76) (1935). *Ranunculus ficaria* Linn. Life-history and pollination. *J. Linn. Soc. Bot. 50* : 39–55.

(77) MARSDEN-JONES, E. M., and TURRILL, W. B. Researches on *Silene maritima* and *S. vulgaris*, parts I to XXV. *Kew Bull.* 1928–47 (continuation in press).

(78) Reports on the transplant experiments of the British Ecological Society at Potterne, Wilts. *J. Ecology, London, 18* : 352–78 (1930) ; *25* : 189–212 (1937) ; *21* : 268–93 (1933) ; *26* : 359–89 (1938) ; *23* : 443–69 (1935) ; *33* : 57–81 (1945).

(79) (1929, 1935). Studies in *Ranunculus*, I and III. *J. Genet. Cambridge, 21* : 169–81 ; *ib. 31* : 363–78.

(80) (1930). The history of a tetraploid saxifrage. *J. Genet. Cambridge*, *23* : 83–92.
(81) (1931). Species studies in plants. *Bot. Exch. Club 1930 Report* : 416–20.
(82) (1931). Flower mutations in the primrose. *New Phyt. 30* : 284–97.
(83) (1933). Studies in variation of *Anthyllis vulneraria*. *J. Genet. Cambridge*, *27* : 261–85.
(84) (1934). Further breeding experiments with *Saxifraga*. *J. Genet. Cambridge*, *29* : 245–68.
(85) (1937). Genetical studies in *Centaurea scabiosa* L. and *C. collina* L. *J. Genet. Cambridge*, *34* : 487–95.
(86) (1938). Further interspecific *Saxifraga* hybrids. *J. Genet. Cambridge*, *36* : 431–45.
(87) (1944). Experiments on colour and heterostyly in the primrose, *Primula vulgaris* Huds. *New Phyt. 43* : 130–34.
(88) MATTHEWS, J. R. (1937). Geographical relationships of the British flora. *J. Ecology, London*, *25* : 1–90.
(88a) (1946). Plant life in Britain : its origin and distribution. *J. R. Hort. Soc. London*, *71* : 225–39, 259–73.
(89) METCALFE, C. R. (1939). The sexual reproduction of *Ranunculus ficaria*. *Ann. Bot. N.S. 3* : 91–103.
(90) MICHAELIS, P. (1940). Über reziprok verschiedene Sippen-Bastarde bei *Epilobium hirsutum*. *Z. indukt. Abstamm.- u. VererbLehre*, *78* : 187–222.
(91) MÜNTZING, A. (1926). Ein Art-Bastard in der Gattung *Lamium*. *Hereditas, Lund*, *7* : 215–28.
(92) (1928). Mendelnde Pollenfarbe bei *Lamium hybridum* Vill. *Hereditas, Lund*, *11* : 284–88.
(93) (1930). Über Chromosomenvermehrung in *Galeopsis*-Kreuzungen und ihre phylogenetische Bedeutung. *Hereditas, Lund*, *14* : 153–72.
(94) (1932). Untersuchungen über Periodizität und Saison-Dimorphismus bei einigen annuellen *Lamium*-Arten. *Bot. Not.* 1932 : 155–76.
(95) (1932). Cytogenetic investigations on synthetic *Galeopsis tetrahit*. *Hereditas, Lund*, *16* : 105–54.
(96) (1937). Multiple allels and polymeric factors in *Galeopsis*. *Hereditas, Lund*, *23* : 371–400.
(97) MÜNTZING, A. and G. (1941). Some new results concerning apomixis, sexuality and polymorphism in *Potentilla*. *Bot. Not.* 1941 : 237–78.
(98) MÜNTZING, A. (1941). New material and cross-combinations in *Galeopsis* after colchicine-induced chromosome doubling. *Hereditas, Lund*, *27* : 193–201.
(99) MÜNTZING, A. and G. (1945). The mode of reproduction of hybrids between sexual and apomictic *Potentilla argentea*. *Bot. Not.* 1945 : 49–71.

(100) Müntzing, A. (1945). Hybrid vigour in crosses between pure lines of *Galeopsis tetrahit*. *Hereditas, Lund, 31* : 391–98.

(101) Newton, W. C. F. (1929). The inheritance of flower colour in *Papaver rhoeas* and related forms. *J. Genet. Cambridge, 21* : 389–404.

(102) (1931). Genetical experiments with *Silene otites* and related species. *J. Genet. Cambridge, 24* : 109–20.

(103) Nilsson, N. H. (1937). Ein oktonärer fertiler *Salix*-Bastard und seine Deszendenz. *Hereditas, Lund, 22* : 361–75.

(104) Patton, D. (1923). Variations in the vegetation along the outcrop of the Lawers-Caenlochan schist. *Bot. Exch. Club 1922 Report* : 797–807.

(105) (1924). The vegetation of Beinn Laoigh. *Bot. Exch. Club 1923 Report* : 268–319.

(106) Philp, J. (1934). Note on the cytology of *Saxifraga granulata* L., *S. rosacea* Moench, and their hybrids. *J. Genet. Cambridge, 29* : 197–201.

(107) Poddubnaja-Arnoldi, V. A. (1939). Hybridization between species of the genus *Taraxacum*. *Bull. Soc. Nat. Moscou, 48* : 87–98 (Russian).

(108) Praeger, R. L. (1901). Irish topographical botany. *Proc. R. Irish Acad. 3rd series, 7*.

(109) (1934). The Botanist in Ireland. Dublin, Hodges, Figgis.

(110) Raunkiaer, C. (1934). The Life Forms of Plants and Statistical Plant Geography. Oxford, Clarendon Press.

(111) Rayner, M. C. (1945). Trees and Toadstools. London, Faber.

(112) Reid, C. (1899). The Origin of the British Flora. London, Dulau.

(113) (1913). Submerged Forests. Cambridge, University Press.

(114) Reid, E. M., and Chandler, M. E. J. (1933). The London Clay Flora. London, British Museum (Natural History).

(115) Ridley, H. N. (1930). The Dispersal of Plants throughout the World. Ashford, Reeve.

(116) Roy, B. (1937). Chromosome numbers in some species and hybrids of *Centaurea*. *J. Genet. Cambridge, 35* : 89-95.

(117) Salisbury, E. J. (1920). The significance of the calcicolous habit. *J. Ecology, London, 8* : 202–15.

(118) (1925). The structure of woodlands. *Veröff. Geobot. Inst. Rübel, 3* : 334–54.

(119) (1925). The incidence of species in relation to soil reaction. *J. Ecology, London, 13* : 149–60.

(120) (1925). Note on the edaphic succession in some dune soils with special reference to the time factor. *J. Ecology, London, 13* : 322–28.

(121) (1932). The East Anglian flora. *Trans. Norfolk Norw. Nat. Soc. 13* : 191–263.

(122) SALISBURY, E. J., and JANE, F. W. (1940). Charcoals from Maiden Castle and their significance in relation to the vegetation and climatic conditions in prehistoric times. *J. Ecology, London, 28*: 310–25.

(123) SANSOME, F. W. (1938). Sex determination in *Silene otites* and related species. *J. Genet. Cambridge, 35*: 387–96.

(124) SAUNDERS, E. R. (1911). An inheritance of a mutation in the common foxglove (*Digitalis purpurea*). *New Phyt. 10*: 47–63.

(125) —— (1918). On the occurrence, behaviour and origin of a smooth-stemmed form of the common foxglove (*Digitalis purpurea*). *J. Genet. Cambridge, 7*: 215–28.

(126) SCOTT, D. H. (1909). Studies in Fossil Botany. London, Black.

(127) —— (1924). Extinct Plants and Problems of Evolution. London, Macmillan.

(128) SEWARD, A. C. (1898–1918). Fossil Plants. Cambridge, University Press. 4 vols.

(129) —— (1931). Plant Life through the Ages. Cambridge, University Press.

(130) SKALINSKA, M. (1944). Polyploidy in *Valeriana officinalis* Linn. in relation to ecology and distribution. *Proc. Linn. Soc. London*, 23 November 1944.

(131) SKENE, M. (1924). The Biology of Flowering Plants. London, Sidgwick & Jackson.

(132) SMITH, K. M. (1943). Beyond the Microscope. Harmondsworth, Penguin Books.

(133) STANFIELD, J. F. (1937). Certain physico-chemical aspects of sexual differentiation in *Lychnis dioica*. *Amer. J. Bot. 24*: 710–19.

(134) STAPF, O. (1914). The southern element in the British flora. *Engl. Bot. Jb. 50 (Suppl.)*: 509–25.

(135) —— (1917). A cartographic study of the southern element in the British flora. *Proc. Linn. Soc. London, 129th Session*: 81–92.

(136) STILES, W. (1936). An Introduction to the Principles of Plant Physiology. London, Methuen.

(137) STOUT, A. B. (1925). Studies of *Lythrum salicaria*:—II. *Bull. Torr. Bot. Club, 52*: 81–85.

(138) SUMMERHAYES, V. S. (1941). The effect of voles (*Microtus agrestis*) on vegetation. *J. Ecology, London, 29*: 14–48.

(139) TÄCKHOLM, G. (1922). Zytologische Studien über die Gattung *Rosa*. *Acta Horti Bergiani, 7*: 97–381.

(140) TANSLEY, A. G. (1923). Practical Plant Ecology. London, Allen & Unwin. (Republished 1946 under the title Introduction to Plant Ecology.)

(141) —— (1939). The British Isles and their Vegetation. Cambridge, University Press.

(142) —— (1945). Our Heritage of Wild Nature. Cambridge, University Press.

(143) TROW, A. H. (1912). On the inheritance of certain characters in the common groundsel—*Senecio vulgaris* Linn.—and its segregates. *J. Genet. Cambridge, 2* : 239–76.

(144) —— (1916). On the number of nodes and their distribution along the main axis in *Senecio vulgaris* and its segregates and on albinism in *Senecio vulgaris* L. *J. Genet. Cambridge, 6* : 1–74.

(145) TURESSON, G. (1943). Variation in the apomictic microspecies of *Alchemilla vulgaris* L. *Bot. Not.* 1943 : 413–27.

(146) TURNER, J. S., and WATT, A. S. (1939). The oakwoods (*Quercetum sessiliflorae*) of Killarney, Ireland. *J. Ecology, London, 27* : 202–33.

(147) TURRILL, W. B. (1928). The flora of St. Kilda. *Bot. Exch. Club 1927 Report* : 428–44.

(148) —— (1929). The flora of Foula. *Bot. Exch. Club 1928 Report* : 838–50.

(149) —— (1931). Biological races in seed-bearing plants and their significance in evolution. *Ann. Appl. Biol. Cambridge, 18* : 442–50.

(150) —— (1936). *Solanum dulcamara* and its inflorescence. *Bot. Exch. Club 1935 Report* : 82–89.

(151) —— (1938). Problems of British *Taraxaca*. *Proc. Linn. Soc. London, 150th Session* : 120–24.

(152) —— (1938). Material for a study of taxonomic problems in *Taraxacum*. *Bot. Exch. Club 1937 Report* : 570–89.

(153) —— (1942). Taxonomy and phylogeny. *Bot. Rev. 8* : 247–70, 473–532, 655–707.

(154) —— (1945). A variety of *Pulicaria dysenterica* from Oxfordshire. *Naturalist, London,* 1945 : 51–52.

(155) —— (1946). The Ecotype Concept. *New Phyt. 45* : 34–43.

(156) UPHOF, J. C. T. (1938). Cleistogamic flowers. *Bot. Rev. 4* : 21–49.

(157) —— (1941). Halophytes. *Bot. Rev. 7* : 1–58.

(158) VALENTINE, D. H. (1941). Variation in *Viola riviniana* Rchb. *New Phyt. 40* : 189–209.

(159) Various Authors (1935). Discussion on the origin and relationship of the British flora. *Proc. Roy. Soc. London, B. 118* : 197–241.

(160) WAHL, H. A. (1940). Chromosome numbers and meiosis in the genus *Carex*. *Amer. J. Bot. 27* : 458–70.

(161) WALKER, A. O. (1912). The distribution of *Elodea canadensis* Michaux in the British Isles in 1909. *Proc. Linn. Soc. London,* 124th Session : 71–77.

(162) WARMKE, H. E., and BLAKESLEE, A. F. (1940). The establishment of a 4n dioecious race in *Melandrium*. *Amer. J. Bot. 27* : 751–62.

(163) WATSON, H. C. (1847–59). Cybele Britannica. London, Longmans. 4 vols. (Supplement, 1860 ; Compendium, 1870).

(164) (1873–74). Topographical Botany. London, Ditton ; 2nd edition 1883, Quaritch.

(165) WEISS, F. E. (1927). Some recent advances in our knowledge of inheritance in plants. *Mem. Manchr. Lit. Phil. Soc. 71 :* 75–86 (*Ranunculus auricomus*).

(166) WILCOX, H. A. (1933). The Woodlands and Marshlands of England. London, Hodder & Stoughton.

(167) WILKINSON, J. (1944). The cytology of *Salix* in relation to its taxonomy. *Ann. Bot. N.S. 8 :* 268–84.

(168) WILLIAMS, R. D. (1937). Genetics of red clover and its bearing on practical breeding. *4th Int. Grassland Congress :* 238–50.

(169) (1939). Genetics of chlorophyll deficiencies in red clover (*Trifolium pratense* L.). *J. Genet. Cambridge, 37 :* 441–58.

(170) (1939). Genetics of cyanogenesis in white clover (*Trifolium repens*). *J. Genet. Cambridge, 38 :* 357–65.

(171) (1941). Incompatibility alleles in *Trifolium pratense* L. ; their frequency and linkage relationships. *Proc. 7th Int. Genet. Congress :* 316.

(172) WILMOTT, A. J. (1930). Concerning the history of the British flora. *Société de Biogéographie, 3.*

(173) WILSON, A. (1931). The altitudinal range of British plants. Suppl. to *North-Western Naturalist*, 1931.

(174) WINGE, Ö. (1931). X- and Y-linked inheritance in *Melandrium. Hereditas, Lund, 15 :* 127–65.

(175) (1932). Experiments with *Papaver rhoeas* L. f. *strigosum* Boenn. *Acad. Sci. SSSR. Bull. Genet. No. 9 :* 115–20.

(176) (1940). Taxonomic and evolutionary studies in *Erophila* based on cytogenetic investigations. *C. R. Lab. Carlsberg, Copenh. Sér. Physiol. 23 :* 41–74.

(177) WOODGER, J. H. (1929). Biological Principles. London, Kegan Paul.

(178) WORSDELL, W. C. (1915–16). The Principles of Plant Teratology. London, Ray Society. 2 vols.

(179) ZEUNER, F. E. (1935). The origin of the English Channel. *Discovery, Cambridge, 16 :* 196–99.

(180) (1945). The Pleistocene Period. London, Ray Society.

(181) (1946). Dating the Past. London, Methuen.

ADDENDUM

(182) SALISBURY, E. J. (1942). The Reproductive Capacity of Plants. London, Bell.

GLOSSARY

abaxial : away from the axis.
absorption : the taking-in of water and solutions.
actinomorphic : regular, so that a cut in one of several planes makes two corresponding halves.
adaxial : towards the axis.
allelomorph : alternative genes or the characters for which they are genetically responsible.
allopolyploid : a polyploid produced by multiplication of dissimilar sets of chromosomes.
allotetraploid : a tetraploid with dissimilar sets of chromosomes.
amphimixis : reproduction by normal fertilization.
androecium : the male part of a flower ; the sum-total of stamens in a flower.
annual : completing a life-cycle within a year.
anther : the head of a stamen, producing the pollen.
anthocyanin : red to blue colouring matter dissolved in the cell-sap.
apetalous : without petals.
apomixis : the production of viable seed without fertilization.
aril : an expansion of the ovule (or seed) stalk.
asexual : without sex.
autecology : the study of the " home-life " of a plant as an individual or as a species.
autopolyploid : a polyploid produced by multiplication of one set of chromosomes.
axil : (of a leaf) the angle made by a leaf with the stem bearing it.
bacteriophages : agents attacking and destroying living bacteria.
bast : tissue of essentially elongated cells with cellulose walls and serving to conduct manufactured foods ; technically termed phloem.
berry : a fleshy fruit, with the seeds embedded in a juicy pulp.
biennial : with a life-cycle extending from one year to the next.
biotic : to do with living organisms.
biotype : a true-breeding group within a larger group.
bisexual : two-sexed.
bivalent : a chromosome pair, associated previously to reduction division.
bract : a modified leaf subtending a flower or an inflorescence branch.
calcareous : limy.
calcicole : living in soil or substratum with high lime-content.
calcifuge : naturally growing only in soil or substratum low in lime-content.

calyx : the outermost whorl of floral leaves.
carbon assimilation : see photosynthesis.
carpel : the unit " female organ," usually consisting of ovary and stigma, with or without a style.
caruncle : a protuberance near the attachment point of a seed.
chalaza : the part of an ovule opposite the mouth (micropyle).
character : a unit abstracted for descriptive purposes.
chasmogamy : the production of flowers which expand for pollination.
chlorophyll : the green colouring-matter of plants.
chromatid : a daughter chromosome, one of the products of division of a chromosome.
chromosomes : components of the nucleus, distinct at stages of nuclear division ; so called from their deep staining with certain dyes.
cleistogamy : the production of unopened flowers in which self-fertilization occurs by pollen-tubes growing from the anthers to the stigma and thence through the style to the ovules in the ovary.
climax : the relatively stable kind of vegetation reached in succession (q.v.) ; by some authors limited to the (relatively) final form of vegetation reached under given climatic conditions.
clone : the sum-total of physiologically independent individuals produced from one sexually produced individual by vegetative propagation or, in extreme usage, without intervention of fertilization.
colchicine : an alkaloid present in autumn crocus (*Colchicum*).
conduplicate : folded together lengthwise.
corolla : the second whorl of floral leaves.
cortex : the part of a stem or root immediately within the superficial layer of cells.
cryptogams : plants not reproducing by seeds.
cuticle : a layer on the outer walls of cells in some organs, contains fatty materials making the walls impermeable to water or water vapour.
cyanogenesis : the production of prussic acid (hydrocyanic acid).
cytogenetics : the study of the microscopic structure of cells combined with study of inheritance.
cytology : the study of the microscopic structure and behaviour of cells.
decaploid : having ten sets of chromosomes.
dioecious : with only androecium or only gynoecium in all flowers of a plant.
diploid : having two sets of chromosomes.
dispersal : the scattering of seeds or other disseminules.
disseminule : a portion of a plant detached and serving to multiply the plant.
distribution : the sum-total of localities occupied by a taxonomic unit.
dorsal : to or on the back.
drupe : a fleshy fruit, with a hard stone ; skin, flesh, and stone being ripened layers of the ovary wall.

ecology : the study of the "home-life" of plants ; of plants in relation to their environments.

ecotype : a group of plants within a species adapted genetically to a limited habitat-range.

ectotrophic : living outside cells.

edaphic : to do with the soil.

embryo : a young organism before separation or emergence from the parent or parental organ.

embryo-sac : the part of an ovule containing the egg-cell and in which the embry normally develops.

enation : a surface outgrowth.

endosperm : food reserve tissue (in a seed) produced inside the embryo-sac from the cell of the fertilized secondary nucleus.

endotrophic : living inside cells.

ephemeral : a short-lived plant with life-history of (at most) a very few months.

epidermis : the outermost layer of cells of leaves and young stems.

evolution : changes in inheritance of organisms such that new kinds are produced from pre-existing kinds.

exotic : foreign ; for this work, not a native of the British Isles.

fasciation : the growth in union of organs normally separate, or the structure resulting from this.

fastigiate : lateral branches more or less parallel with the main stem or trunk.

fertilization : the fusion of male and female sex cells (gametes).

filament : the stalk of a stamen.

flavone : a yellow colouring matter in plastids.

flora : the kinds of plants (as species, etc.) occurring in an area ; sometimes extended to mean a publication dealing with the flora of an area ; hence floristics.

fluctuation : the modification of a character as a direct reaction to environmental factors.

fruit : technically the ripened ovary or ovaries after fertilization ; often extended to include other persistent and ripening parts of the flower, receptacle, infructescence, etc.

gamete : a sex cell (male or female).

gene : the unit of inheritance.

genetics : the study of inheritance (hence genetical, geneticist).

genotype : the breeding potentialities of an organism and by extension a group of organisms of similar breeding potentialities.

germination : the early stages of growth of the embryo in a seed, or of a spore.

glabrous : without hairs.

glucosides : organic chemical compounds which yield a sugar (glucose) and other substances on breaking down.

gynodioecism : having hermaphrodite and female flowers on different individuals.
gynoecium : the female part of a flower ; that containing the ovules and associated parts concerned in pollen reception and fertilization.
habitat : the sum-total conditions under which a plant lives.
halophyte : a plant adapted to living under saline conditions (usually of the soil or soil water).
haploid : having one set of chromosomes.
hermaphrodite : with androecium and gynoecium.
heterostyly : having more than one length of style and, usually, differences in length of filaments or position of anthers in the flowers of different plants of one species.
heterozygote : a hybrid, at least for given characters or genes ; hence heterozygous.
hexaploid : having six sets of chromosomes.
homostyle : having only one style-length in flowers of different plants ; also used when stigma (or stigmata) are level with the anthers.
homozygote : pure, non-hybrid, similar genes in homologous loci, at least for given genes ; hence homozygous.
humus : organic material in the soil produced by the decay of dead bodies (or organs) of plants and animals.
hybrid : a cross between two parents with different genes ; sometimes in practice restricted to crosses between species or well-marked varieties.
hydrophyte : a plant living more or less in water.
hygrophyte : a plant living under marshy conditions.
hypha (pl. *hyphae*) : a fine thread of living cells.
hypocotyl : the stem of an embryo or seedling below the seed-leaves (or cotyledons).
incompatibility : used technically to denote exceedingly slow, and usually unsuccessful, growth of pollen-tubes through stylar tissue.
indumentum : hairy covering.
inflorescence : a more or less clearly defined grouping of flowers.
infructescence : a more or less clearly defined grouping of fruits.
intergeneric : between two (or more) genera.
interspecific : between two (or more) species.
intraspecific : within one species.
karyology : the study of the structure and behaviour of nuclei.
lamina : a leaf-blade.
lenticel : a special breathing-pore of woody organs.
lethal genes : genes responsible for the early death of sex cells or plantlets.
locus : used technically to designate the position of a gene in a chromosome.
meiosis : the process of cell divisions in which the chromosome number is

reduced, typically from diploid to haploid, previous to formation of gametes or sex cells.

mesophyte : a plant living under neither very wet nor very dry conditions.

metamorphosis : change in the structure of an organ so that it resembles some other organ.

microclimate : climatic conditions within a very small area, and especially within a plant community or a portion thereof.

microevolution : small-scale evolution ; the evolving of new varieties and species in contrast to major evolutionary trends.

mitosis : the process of division in an ordinary body-cell.

monoecious : with androecium or gynoecium in separate flowers but both kinds of flowers on the same plant.

monohybrid : with only one pair of characters (or genes) considered in the hybrid condition or in segregation from such.

morphology : the study of the structure or form of organisms.

multicellular : of several or many cells.

multiple allelomorphs : a series of gene mutations at one locus (position) on a chromosome.

mutation : a definite change in a character, passed on to future generations by changes in reproductive cells.

mycorrhiza : root specialized through association with a fungus.

nucleus (pl. *nuclei*) : a specialized portion of protoplasm, directive of many cell activities and essential in reproduction and inheritance.

octoploid : having eight sets of chromosomes.

ontogeny : the life-history of an individual or of a part (organ) of an individual.

ovary : the portion of the female part of a flower enclosing the ovules.

ovule : a structure which can, given suitable fertilization and conditions, develop into a seed.

paramorph : any taxonomic variant within a species ; conveniently used when the evidence is insufficient to define the status more precisely.

parasites : organisms living upon living organisms.

parthenogenesis : the development of an egg-cell (female gamete) without fertilization.

pedicel : a flower-stalk.

peduncle : an inflorescence-stalk.

peloria : the production of regular in place of irregular flowers.

pentaploid : having five sets of chromosomes.

perennial : living three or more years.

perianth : the calyx and corolla considered without distinction.

perisperm : food reserve tissue (in a seed) formed from that part of the ovule in which the embryo-sac is embedded.

petaloid : petal-like.

petals : the individual members of the corolla.

petiole : a leaf-stalk.
phanerogams : plants reproducing by seeds (spermatophytes is a better term).
phenology : the study of recurring seasonal behaviour.
phenotype : the demonstrable characters of an organism and by extension a group of organisms of similar demonstrable characters, apart from breeding potentialities.
phloem : see "bast."
photosynthesis : the manufacture of food by chlorophyll from water and carbon dioxide through utilization of sunlight (also termed carbon assimilation).
phyllody : becoming like foliage leaves.
phylogeny : the origin and course of evolution of a systematic group.
physiology : the study of the activities, behaviour, functions of organisms.
placenta : the part of an ovary to which ovules are attached, or of a fruit to which seeds are attached.
plant community : any grouping of plant individuals in nature.
plastid : a specialized portion of protoplasm, holding colouring matter or connected with starch formation and storage.
pollen : the grains containing, in seed-bearing plants, the male sex-cells (male gametes).
pollination : the transfer of pollen from anther to stigma.
pollinium (pl. *pollinia*) : a mass (often club-shaped) of pollen and grains adhering together and transported as a unit in pollination.
polymeric : represented several times (as genes of the same type).
polyploid : having more than two sets of chromosomes in the nuclei of the body cells.
polysomic : the condition of a gene represented several times in a nucleus because of chromosome multiplication.
pome : a fleshy fruit, with ripened receptacle and ovary wall surrounding a core with the seeds.
primordium (pl. *primordia*) : the earliest beginning of an organ.
propagule : any portion of a plant which multiplies the number of physiologically independent individuals.
prothallium : a flat green structure which is the sexual generation in ferns and fern-allies.
protoplasm : the living material of plant cells.
raceme : an inflorescence with stalked flowers arising in continuous succession on an elongated axis with the oldest flowers below.
ramet : a single physiologically independent individual of a clone (q.v.).
range : the geographical area over which a taxonomic unit extends.
raphe : a vascular bundle running down one side of a bent-over ovule.
ray florets : outer, usually strap-shaped, florets in a compact inflorescence.
receptacle : the portion of a flower-stalk bearing or surrounding the floral organs.

reduction division : the double division at which chromosome numbers are halved.

respiration : the breakdown of organic food in living tissues by oxidation with the release of carbon dioxide and energy.

reticulate : with a network structure, arrangement, or markings.

rhizome : an elongated, usually more or less horizontal, underground stem.

ruderal : a plant living in waste places, such as path-sides, rubbish-heaps, bases of dwellings.

saprophytes : plants living on dead organic matter.

seed : a detachable reproductive structure containing a young plant (or embryo).

segregation : the separation of contrasted characters in individuals of a family.

self-fertilization : fusion of male and female sex cells produced by the same individual.

self-pollination : transfer of pollen from anther to stigma in the same individual plant.

sepaloid : sepal-like.

sepals : the individual members of the calyx.

sere : an actual example of succession (q.v.) ; hence seral.

sex : the total of reproduction by fusion of special cells.

somatic : body cells, as distinct from sex cells.

spermatophytes : seed-bearing plants.

spike : an inflorescence with sessile flowers arising in continuous succession on an elongated axis with the oldest flowers below.

sporangium (pl. *sporangia*) : an organ giving rise to spores internally.

spore : a definite reproductive unit of a simple kind, consisting of one or a few cells and not differentiated into organs or parts.

sporophylls : special leaves associated with spore-bearing organs.

stamen : the structure in seed-bearing plants which produces the pollen-grains.

stigma (pl. *stigmata*) : the portion of the female part of the flower receptive to pollen.

stoma (pl. *stomata*) : a special breathing-pore of leaves and young stems.

style : a prolongation of the top of an ovary bearing the surface receptive to pollen.

succession : used technically for the series of changes shown by vegetation in correlation with habitat changes.

synecology : the study of the " home-life " of plant communities.

taxonomy : the study of plant classification.

tepals : the individual members of the perianth.

tetrad : a group of four.

tetraploid : having four sets of chromosomes.

tetrasomic : with genes or chromosomes represented four times.
tiller : a side shoot (particularly in grasses).
transpiration : the loss of water vapour from plant organs.
triploid : having three sets of chromosomes.
tristyly : three different style-lengths within the species.
unicellular : of one cell.
unisexual : of one sex only.
univalent : an unpaired chromosome at the reduction division.
vascular bundles : the food- and water-conducting strands in plants.
vegetation : plants grouped together under certain living conditions, irrespective of their systematic (or floristic) groupings.
venation : the system of veins, as in a leaf.
ventral : to or on the front.
vernation : the arrangement of leaves in the bud.
viruses : disease-causing substances produced by living cells and passing through fine filters.
wood : tissues of elongated cells with walls impregnated with " lignin " and serving to conduct water and solutions ; technically termed xylem.
xerophyte : a plant living under dry conditions.
xylem : see " wood."
zygomorphic : irregular, so that a division into two equivalent halves can be made in one plane only, as in a longitudinal cut through a pea flower.
zygote : the cell resulting from the fusion of a male and female sex-cell at fertilization.

INDEX

Numbers in parentheses usually are of pages where a plant is referred to under a Latin but not under an English name or vice versa, or, more rarely, where an alternative vernacular name is used. Numbers in heavy type refer to pages opposite which illustrations will be found.